BIOENERGETICS

BIOENERGETICS

An Introduction to the Chemiosmotic Theory

DAVID G. NICHOLLS

Neurochemistry Laboratory
Department of Psychiatry
Ninewells Medical School
University of Dundee
Dundee
Scotland

1982

ACADEMIC PRESS

A Subsidiary of Harcourt Brace Jovanovich, Publishers

LONDON   NEW YORK
Paris   San Diego   San Francisco
São Paulo   Sydney   Tokyo   Toronto

ACADEMIC PRESS INC. (LONDON) LTD.
24/28 Oval Road,
London NW1

United States Edition published by
ACADEMIC PRESS INC.
111 Fifth Avenue
New York, New York 10003

British Library Cataloguing in Publication Data

Nicholls, D. G.
Bioenergetics
I. Title
574.19′121   QH510

ISBN 0–12–518120–5   Casebound edition
ISBN 0–12–518122–1   Paperback edition

Preface

Bioenergetics has existed as a distinct discipline within biochemistry for the past thirty years. However, despite the great rationalization of the field which has stemmed from the chemiosmotic hypothesis of Peter Mitchell, it is still difficult for the student or the non-specialist biochemist to follow the bioenergetics research literature. To a great extent this problem is a consequence of the large gap which exists between the general biochemistry textbooks, which for reasons of space are constrained to treat the subject superficially, and the research literature, which tends to be highly technical, usually contentious, and sometimes obscure.

This book is intended as an introduction to the field rather than a comprehensive review. First therefore, although dissenting opinions exist for almost every topic covered, I have based the contents on what I feel to be the current orthodoxy in bioenergetics. In particular I have assumed from the outset the "central dogma" of the chemiosmotic hypothesis: energy transduction by a proton electrochemical gradient. Secondly, I have been highly selective in the use of references and have only cited easily accessible reviews together with a limited number of original papers for purposes of illustration. These reviews allow access to the complete bioenergetics bibliography, thus allowing the book to be shortened considerably. I hope, however, that colleagues will recognize that the selection of original papers is in no way intended to reflect scientific precedence.

I have presupposed a general knowledge of biochemistry equivalent to that available from current student textbooks of general biochemistry, and so the book is intended for final year undergraduates, graduate students, and non-specialist research workers.

Thanks are due to Abraham Tulp and TAB for the cartoons. Those prefacing Chapters 2, 4, 5, 6 and 7 are reproduced by permission of *Trends in Biochemical Sciences*.

Dundee, September 1981   *David G. Nicholls*

Contents

1 Chemiosmotic energy transduction   **1**

  1.1   What is "energy transduction"?   1
  1.2   The chemiosmotic hypothesis   3
  1.3   Energy-transducing organelles   5
  1.4   Background to the chemiosmotic hypothesis   12

2 Ion transport across energy-transducing membranes   **25**

  2.1   Introduction   25
  2.2   The structure of energy-transducing membranes   25
  2.3   Pathways of ion transport   26
  2.4   The natural permeability properties of bilayer regions   29
  2.5   Ionophore-induced permeability properties of bilayer regions   30
  2.6   Protein-catalysed transport   35
  2.7   Bulk solute movements across energy-transducing membranes   36

3 Quantitative bioenergetics: the measurement of driving forces   **41**

  3.1   Introduction   41
  3.2   Gibbs energy   44
  3.3   Oxido–reduction (redox) potentials   49
  3.4   Ion electrochemical potential differences   53
  3.5   Equilibrium distributions of ions, weak acids and weak bases   56
  3.6   Membrane potentials, diffusion potentials, Donnan potentials and surface potentials   58
  3.7   Photons   59
  3.8   Bioenergetic interconversions   60
  3.9   The application of irreversible thermodynamics   62

4 The chemiosmotic proton circuit   **65**

  4.1   Introduction   65
  4.2   The measurement of proton electrochemical potential   68
  4.3   The stoicheiometry of proton extrusion by the respiratory chain   78
  4.4   The stoicheiometry of proton uptake by the ATP synthetase   82
  4.5   Proton current, proton conductance and respiratory control   84
  4.6   Non-chemiosmotic parameters of energy-transduction   91
  4.7   Reversed electron transfer and the proton circuit driven by ATP hydrolysis   93
  4.8   ATP synthesis driven by an artificial proton electrochemical potential   95

5   Respiratory chains   **99**

   5.1   Introduction   99
   5.2   Components of the mitochondrial respiratory chain and
         methods of detection   99
   5.3   The linear sequence of redox carriers in the respiratory
         chain   108
   5.4   Proton translocation by the respiratory chain; structural
         predictions made by "loop" or "conformational pump"
         models   114
   5.5   Fractionation and reconstitution of respiratory chain
         complexes   115
   5.6   Complex I (NADH–UQ oxidoreductase)   117
   5.7   Complex II (succinate dehydrogenase); electron-
         transferring flavoprotein and α-glycerophosphate dehy-
         drogenase   118
   5.8   Ubiquinone and complex III (bc_1-complex or UQ-cyt c
         oxidoreductase)   119
   5.9   Cytochrome c and complex IV (cytochrome c oxidase;
         ferrocytochrome c-O_2 oxidoreductase   122
   5.10 The nicotinamide nucleotide transhydrogenase   124
   5.11 The respiratory chain of plant mitochondria   126
   5.12 Bacterial respiratory chains   127

6   Photosynthetic generators of proton electrochemical potential   **133**

   6.1   Introduction   133
   6.2   The light reaction of purple bacteria   135
   6.3   The generation of $\Delta\tilde{\mu}_{H+}$ in purple bacteria   138
   6.4   The electron-transfer pathway in chloroplasts   142
   6.5   The proton circuit in chloroplasts   146
   6.6   Bacteriorhodopsin and the purple membrane of
         halobacteria   147

7   The ATP synthetase   **151**

   7.1   Introduction   151
   7.2   The structure of the ATP synthetase   151
   7.3   The function of F_0 154
   7.4   The mechanism of ATP synthesis by F_1 155
   7.5   The transport of adenine nucleotides and Pi in
         mitochondria   159

8   The interaction of bioenergetic organelles with their environment   **167**

   8.1   Introduction   167
   8.2   Methods of studying metabolic transport   167
   8.3   Mitochondrial metabolite carriers   169
   8.4   Mitochondrial calcium transport   172
   8.5   Bacterial transport   175

   References   179
   Subject index   185

Glossary of Common Symbols and Abbreviations

Ac	Acetate
AcAc	Acetoacetate
ADP/O	The number of molecules of ADP phosphorylated to ATP when two electrons are transferred from a substrate through a respiratory chain to reduce one "O" ($\frac{1}{2}O_2$) to water (dimensionless) (Section 4.6)
$ADP/2e^-$	As ADP/O except more general as the final electron acceptor can be other than "O" (dimensionless) (Section 4.6)
Bchl	Bacteriochlorophyll
Bpheo	Bacteriopheophytin
CF_1	(see F_1)
C-face	The face of the inner mitochondrial membrane in contact with the cytosol (cf. M-face)
Chl	Chlorophyll
C_MH^+	The effective proton conductance of a membrane or a membrane component (dimensions: nmol H^+ min^{-1} mg $protein^{-1}$ mV of proton electrochemical $potential^{-1}$) (Section 4.5)
DNP	Dinitrophenol (proton translocator)
DCCD	Dicyclohexylcarbodiimide (ATP synthetase inhibitor) (Section 7.2)
dO/dt	Respiratory rate (dimensions: nmol O min^{-1} mg $protein^{-1}$)
E	Redox potential at pH$=0$ (dimensions: mV) (Section 3.3)
E_0	Standard redox potential at pH$=0$ (dimensions: mV) (Section 3.3)
E_h	Actual redox potential at a defined pH (dimensions: mV) (Section 3.3)
E_m	Mid-point potential (standard redox potential at a defined pH) (dimensions: mV) (Section 3.3)
$E_{h,7}$	Actual redox potential at pH$=7$
$E_{m,7}$	Mid-point potential at pH$=7$
ETP, etp	Electron transfer particle (sub-mitochondrial particle) (Section 1.3)
F	The Faraday constant ($=0\cdot0965$ kJ mol^{-1} mV^{-1}); to convert from mV to kJ $mole^{-1}$, multiply by 'F' (Section 3.3)
FCCP	Carbonyl cyanide-p-trifluoromethoxyphenylhydrazone, a proton translocator
Fe/S centre	The prosthetic group of a class of redox proteins containing acid-labile non-haem iron (Section 5.2)
F_0	A subunit of the ATP synthetase (Section 7.3)
F_1, CF_1, TF_1	The catalytic subunit of the ATP synthetases from, respectively, mitochondria, chloroplasts and thermophilic bacteria (Section 7.3)

G	Gibbs energy content (Section 3.2)
H	Enthalpy content
H^+/ATP	The number protons translocated through the ATP synthetase for the synthesis of one molecule of ATP (dimensionless) (Section 4.4)
H^+/O	The number of protons translocated by the respiratory chain during the passage of 2 e^- from substrate to oxygen (dimensionless) (Section 4.3)
$H^+/2e^-$	As H^+/O except more general as final electron acceptor need not be oxygen (dimensionless) (Section 4.3)
$h\nu$	The energy in a photon (dimensions: kJ) (Section 3.7)
J_{H+}	Proton current (dimensions: nmol H^+ min^{-1} mg protein^{-1}) (Section 4.5)
K	Absolute equilibrium constant (Section 3.2)
K'	Apparent equilibrium constant (Section 3.2)
kD	Kilodalton: units of 1000 molecular weight
kJ	Kilojoule
M-face	The face of the inner mitochondrial membrane in contact with the matrix
NEM	N-ethylmaleimide
O	$\frac{1}{2}O_2$
OSCP	Oligomycin sensitivity conferring protein (Section 7.2)
PC	Plastocyanine (Section 6.4)
PEP	Phosphoenolpyruvate
Pi	orthophosphate
PMS	Phenazinemethosulphate
PQ	Plastoquinone
P/O	As ADP/O
$P/2e^-$	As ADP/$2e^-$
PS_I, PS_{II}	Chloroplast photosystems (Reaction centres) I and II (Section 6.4)
P_{870} etc.	The primary photochemical component in a reaction centre (Section 6.2)
$q^+/0$	The number of charges transduced across the membrane when 2 e^- are transferred from substrate to oxygen through the respiratory chain (Section 4.3)
$q^+/2e^-$	As $q^+/0$ but more general as final electron acceptor need not be oxygen (Section 4.3)
R	The gas constant ($=0.0083$ kJ mole^{-1} K^{-1})
RC	Reaction centre (Section 6.1)
SMP	Sub-mitochondrial particle
S	Entropy
T	Absolute temperature in degrees Kelvin (K)
TMPD	Tetramethyl-p-phenalinediamine. Redox mediator (Section 5.9)
$TPMP^+$	Triphenylmethyl phoshonium cation: electrically permeant synthetic cation (Section 2.5)
$UQ, U\dot{Q}H, UQH_2$ UQ_2, UQ_8, UQ_{10}	Oxidized, semiquinone and reduced forms of ubiquinone; subscripts refer to number of isoprenoid groups in side-chain

Γ, Γ'	Absolute and apparent observed mass–action ratios (Section 3.2)
ΔE_h	Actual redox potential difference between two couples at a defined pH (dimensions: mV) (Section 3.3)
$\Delta E_{h,7}$	As ΔE_h, but for pH $= 7$ (Section 3.3)
ΔG	The change in Gibbs energy (dimensions: kJ mole^{-1}) (Section 3.2)
ΔG^0	The standard Gibbs energy change based on absolute equilibrium constants (Section 3.3)
$\Delta G^{0'}$	The standard Gibbs energy change based on apparent equilibrium constants; pH etc. must be specified (Section 3.3)
ΔG_p	The "phosphorylation potential", i.e. the Gibbs energy change for the synthesis of ATP from ADP and Pi (Section 3.3)
ΔH	Enthalpy change
ΔpH	The pH difference between the bulk phases on either side of a membrane (dimensionless)
ΔS	Entropy change
$\Delta\tilde{\mu}_{X^{m+}}$	The electrochemical potential difference for the ion X^{m+} between two bulk phases separated by a membrane (Section 3.4)
$\Delta\tilde{\mu}_{H^+}$	The electrochemical potential difference for protons between two bulk phases separated by a membrane (dimensions: mV) (Section 3.4); sometimes called protonmotive force (pmf)
$\Delta\Psi$	Membrane potential, i.e. the electrical potential difference between two bulk phases separated by a membrane (dimensions: mV) (Section 3.4)
"$\sim$"	"Squiggle": the shorthand symbol in the chemical hypothesis for the hypothetical energy-transducing intermediate

 Printed in Great Britain by
Page Bros (Norwich) Ltd

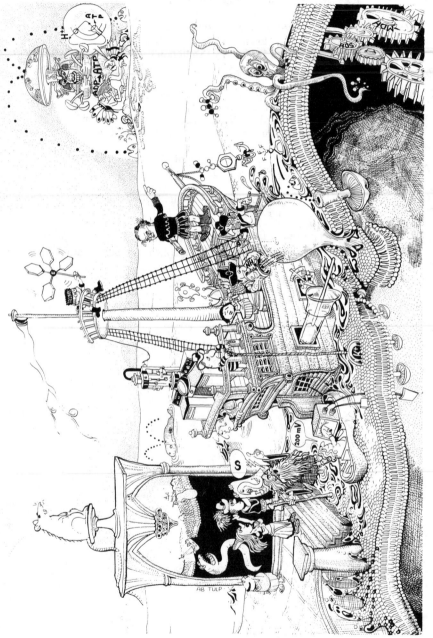

Mitchell sets sail for the Chemiosmotic New World, despite dire warnings that he will be consumed

1 Chemiosmotic Energy Transduction

1.1 WHAT IS "ENERGY TRANSDUCTION"?

All biochemical reactions involve energy changes, so the term "bioenergetics" could be applied to the whole of biochemistry. Bioenergeticists, however, tend to restrict their investigations to a particular type of process which occurs in a distinct class of membrane. The central theme in bioenergetics for the past thirty years has been to understand the mechanisms by which the energy made available by the oxidation of substrates, or by the absorption of light, can be used to drive "uphill" reactions such as the synthesis of ATP from ADP and Pi, or the accumulation of ions across a membrane.

Although some ATP synthesis occurs in soluble enzyme systems, by far the largest proportion is associated with membrane-bound enzyme complexes which are restricted to a particular class of membrane. These "energy-transducing" membranes are the plasma membrane of simple prokaryotic cells such as bacteria or blue-green algae, the inner membrane of mitochondria, and the thylakoid membrane of chloroplasts (Fig. 1.1). The mechanism of ATP synthesis and ion transport by these diverse membranes is sufficiently related, despite the differing natures of their primary energy sources, to form a single field of study: *energy transduction* or *bioenergetics*.

Energy-transducing membranes possess a number of distinguishing features. Each membrane has two distinct protein assemblies. One is usually called the ATPase, but should more correctly be termed the ATP synthetase, and catalyses the "uphill" synthesis of ATP from ADP and Pi. It is common to all energy-transducing membranes. The nature of the second assembly depends on the primary energy source for the membrane; in the case of mitochondria or respiring bacteria it is a respiratory chain catalysing the "downhill" transfer of electrons from substrates to final acceptors such as O_2. In chloroplasts and photosynthetic bacteria the second assembly uses the energy available from the absorption of quanta of visible light (Fig. 1.1).

The identity of the "energy-transducing intermediate" linking these pairs of protein assemblies proved to be highly elusive. Many years of searching for a chemical intermediate to couple oxidation to ATP synthesis was without success, and it reached a stage when uncertainty was so extreme that it

1

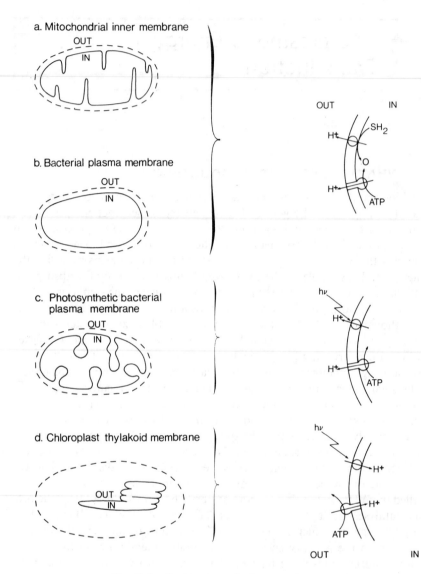

Fig. 1.1 Energy-transducing membranes.

prompted the statement that "anybody who is not thoroughly confused just doesn't understand the problem". It was at this point that Peter Mitchell entered the scene with the chemiosmotic hypothesis (1961) in which he suggested that the only "intermediate" was a proton gradient across the membrane. The subsequent, frequently acrimonious, debate between advocates of chemical, chemiosmotic and other hypotheses continued unabated for 15 years (see Boyer *et al.* 1977). As a result the chemiosmotic hypothesis has probably been submitted to a more rigorous trial than any other comparable biochemical theory. The steady flow of converts to the hypothesis was ultimately recognized with the award to Peter Mitchell of the Nobel Prize in Chemistry for 1978 (see Garland 1978, Chappell 1979).

1.2 THE CHEMIOSMOTIC HYPOTHESIS

Reviews Mitchell 1961, 1966, 1968, 1979a, Greville 1969

The central dogma of the chemiosmotic hypothesis states that the electron-transfer chains of mitochondria, chloroplasts or bacteria are coupled to ATP synthesis by a proton electrochemical potential ($\Delta\tilde{\mu}_{H+}$) across the energy-transducing membrane. Proton electrochemical potential is a thermodynamic measure of the extent to which the proton gradient across the membrane is removed from equilibrium; it will be derived in Section 3.4. Electron transfer and ATP synthesis are catalysed by separate reversible proton pumps. The $\Delta\tilde{\mu}_{H+}$ generated by electron transfer is used to drive an ATP-hydrolysing proton pump (the ATPase, or more correctly the ATP-synthetase) backwards, i.e. in the direction of ATP synthesis.

Each energy-transducing membrane therefore contains two proton pumps, one driven by electron transfer or photon capture and one driven (in the forward direction) by ATP hydrolysis (Fig. 1.1). The two pumps have the same orientation, that is to say the "downhill" transfer of electrons along the electron-transfer chain causes proton translocation across the membrane in the same direction as that caused by the "downhill" hydrolysis of ATP by the ATP-synthetase (Fig. 1.1). Within each proton pump the translocation of protons across the membrane is tightly linked to electron transfer or ATP hydrolysis.

Figure 1.2 explains how these pumps can be used for the continuous synthesis of ATP. If conditions were arranged so that only the ATP synthetase were active and ATP was added to the organelle, the nucleotide would be hydrolysed and protons would be pumped until an equilibrium was established between the energy available from the further hydrolysis of ATP and the energy required to pump further protons against the proton gradient which had already been established (Fig. 1.2b). If this equilibrium is now disturbed by removing ATP, the ATP synthetase would reverse and run in the direction of net ATP synthesis, driven by the proton gradient (Fig. 1.2c). This, however, would rapidly collapse the gradient, and so a second proton pump is required to continuously replenish it. This is precisely what occurs

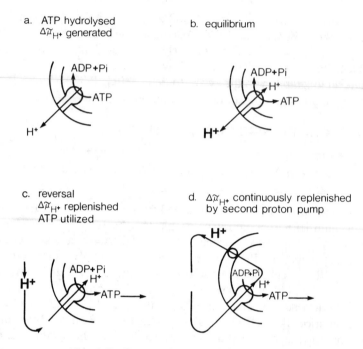

Fig. 1.2 Proton electrochemical coupling.

in vivo: ATP is continuously removed for cytosolic ATP-consuming reactions, while $\Delta\tilde{\mu}_{H+}$ is continuously replenished by the respiratory or photosynthetic electron-transfer chains (Fig. 1.2d).

The combined effect of these proton translocations is to establish a circuit of protons across the energy-transducing membrane (Fig. 1.3). This proton circuit is closely analogous to an electrical circuit, and the analogy holds even when discussing detailed and complex energy flows (see Chapter 4). As with the electrical case one can measure a potential (the proton electrochemical potential, $\Delta\tilde{\mu}_{H+}$ which has already been mentioned), a current of protons (J_{H+}), and a conductance for protons ($C_M H^+$), defined from the current flowing through a component divided by the potential drop across the component. Frequent use of these parameters will be made in later chapters.

To avoid short-circuiting, it is evident that the membrane must be closed and must possess a high resistance to protons. One of the most successful predictions of the hypothesis was that many agents which uncouple oxidation from ATP synthesis act by increasing the proton conductance of the membrane and inducing just such a short-circuit (Fig. 1.3). To allow negatively

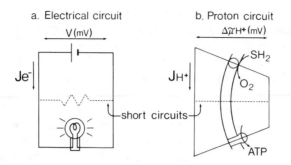

Fig. 1.3 Proton circuits and electrical circuits are analogous. Both have generators of potential (the battery and the respiratory chain respectively); both potentials (voltage difference and proton electrochemical potential difference) are expressed in volts or millivolts. Both potentials can be made to perform work (the light bulb and the ATP synthetase). Both circuits can be short circuited. The rate of chemical conversion in both the battery and respiratory chain is tightly linked to the current of electrons and protons flowing in the rest of the circuit, which in turn depends on the resistance of the rest of the circuit. The potentials fall as the current drawn increases.

charged metabolites to enter the negative interior of a mitochondrion, the chemiosmotic hypothesis also included a proposal for the existence of transport systems in which metabolites were transported as neutral species together with protons, or in exchange for OH^- (Mitchell 1961, 1966).

There is considerable semantic confusion as to what constitutes the "chemiosmotic hypothesis" or the "Mitchell theory". In this book the term "chemiosmosis" will be used synonymously with the central dogma outlined above, namely that coupling occurs through the intermediacy of a proton electrochemical gradient. There are many secondary hypotheses, by Peter Mitchell and others, as to the precise molecular mechanism by which the pumps operate, but it must be emphasized that the validity of the central dogma is independent of the fate of these secondary theories.

1.3 ENERGY-TRANSDUCING ORGANELLES
1.3.1 Mitochondria and sub-mitochondrial particles

Reviews Nedergaard & Cannon 1979 (preparation); Munn 1974 (morphology); Harmon *et al.* 1974, DePierre & Ernster 1977 (location of enzymes)

The appearance of a typical mitochondrion in thin-section is shown in Fig. 1.4. Mitochondria are typically 0·7–1 μm in length. Their shape is not fixed but varies continuously in the cell, and the appearance of the cristae can be quite different in mitochondria isolated from different tissues or even with the same mitochondria suspended in different media.

The outer membrane possesses proteins which act as non-specific pores for solutes of less than 10 kD (Zalman *et al.* 1980); the inner membrane is energy transducing. The greater the respiratory activity of the tissue from which the mitochondria are isolated, the more extensively the inner membrane is infolded into cristae. In mitochondrial preparations which have been negatively stained with phosphotungstate, knobs are visible on the matrix face (M-face) of the inner membrane; these are the catalytic units of the ATP synthetase. The enzymes of the citric acid cycle are in the matrix, except succinate dehydrogenase, which is bound to the M-face of the inner mem-

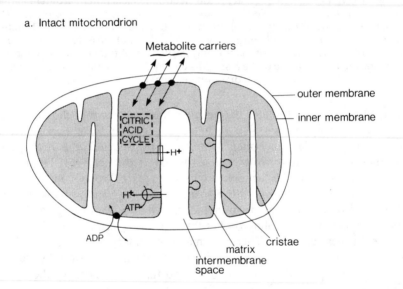

a. Intact mitochondrion

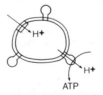

b. Sub – mitochondrial particle

Fig. 1.4 Preparations derived from mitochondria.

brane. The matrix pools of NAD and NADP are separate from those in the cytosol, while matrix ADP and ATP communicates with the cytosol through the adenine nucleotide exchanger.

Mitochondria are usually prepared by gentle homogenization of the tissue in isotonic sucrose, followed by differential centrifugation to separate mitochondria from nuclei, cell debris and microsomes (fragmented endoplasmic reticulum). Although this method is effective with fragile tissues such as liver, tougher tissues such as heart must either be first incubated with a protease, such as nagarse, or be exposed briefly to a blender to break the muscle fibres. Yeast mitochondria are isolated following digestion of the cell wall with snail-gut enzyme.

Ultrasonic disintegration of mitochondria produces inverted submitochondrial particles (SMPs) (Fig. 1.4), also called electron-transport particles (ETPs). Because these have the substrate binding sites for both the respiratory chain and the ATP synthetase on the outside, they are much used (see Lee 1979).

1.3.2 Respiring bacteria and derived preparations

Reviews Kaback 1974, Konings 1979

Although some aspects of bacterial energy transduction can be studied with intact bacteria, the cell wall, particularly of Gram-negative strains, can introduce permeability problems. For this reason sphaeroplasts (Fig. 1.5) which have the cell wall removed by lysozyme digestion are frequently used. Sphaeroplasts retain their contents and orientation but become osmotically sensitive.

If cells are extruded at very high pressure through an orifice in a French press, they invert to form vesicles analogous to sub-mitochondrial particles. Conversely, hypotonic lysis of sphaeroplasts yields empty membrane vesicles ("ghosts") retaining the original polarity. These are known as "Kabackosomes".

1.3.3 Chloroplasts

Review Park & Sane 1971

Chloroplasts are one of a variety of *plastids* or vesicles peculiar to green algae and higher plants; there may be from one to a hundred chloroplasts per cell. Chloroplasts are considerable larger than the average mitochondrion, being

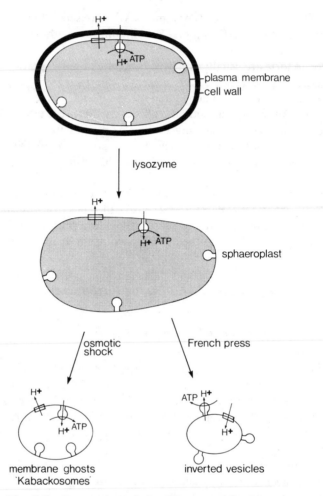

Fig. 1.5 Preparations derived from respiring bacteria.

4–10 μm in diameter and 1–2 μm thick (Fig. 1.6). They are bounded by an envelope of two closely opposed membranes, the matrix within the inner membrane being the *stroma*. Within the stroma are flattened vesicles called thylakoids. Several thylakoids may be stacked together at specific regions to form grana. The thylakoid membranes contain the photosynthetic pigments and the carriers of photosynthetic electron transfer. These are the energy-transducing membranes, and light causes the injection of protons into the internal thylakoid spaces. The chloroplast ATP synthetase is bound to the thylakoid membrane and orientated with its "knobs" on the stromal face of

a. Whole chloroplast

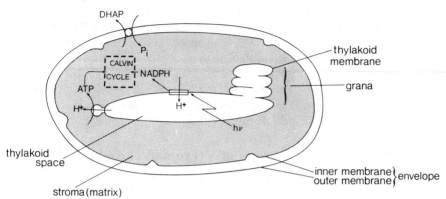

b. Chloroplast lamellar system

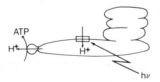

Fig. 1.6 Preparations derived from chloroplasts.
(a) "intact" chloroplast;
(b) "broken" chloroplast—note that the thylakoid membrane is still intact.

the membrane. The ATP and NADPH generated by photosynthetic phosphorylation is used by the CO_2-fixing dark reactions of the Calvin cycle located in the stroma.

Although at first sight the structure of chloroplasts appears to be very different from that of mitochondria, the only topological distinction is that the thylakoids, in contrast to the mitochondrial cristae, have become separated from their respective inner membranes, with the result that the thylakoid space becomes a separate compartment, unlike the "cristal space" which is continuous with the inter-membrane space of mitochondria.

Chloroplasts are prepared by gentle homogenization of leaves in isotonic sucrose or sorbitol at pH 8 to compensate for the rupture of acidic vacuoles.

After filtering through muslin, the suspension is centrifuged slowly to remove debris, and the chloroplasts are then pelleted by centrifugation typically at $1-2000 \times g$ for 4 min. This preparation gives chloroplasts with largely intact inner and outer membranes (intact chloroplasts). These show high rates of CO_2 fixation due to their retention of stromal contents but are of limited use for bioenergetic studies because the membranes prevent many substrates and inhibitors from reaching the thylakoid membrane. Choroplasts prepared in 0.35 M NaCl have damaged inner and outer membranes, do not fix CO_2, but allow access to the thylakoid membrane (broken chloroplasts).

1.3.4 Photosynthetic bacteria and chromatophores

Review Clayton & Sistrom 1979

Three groups of prokaryotes catalyse photosynthetic electron transfer: the green bacteria, the purple bacteria, and the cyanobacteria (or blue-green algae). The purple bacteria are divided into two groups: the Rhodospirillaceae (or non-sulphur) and the Chromatiaceae (or sulphur). Cyanobacteria carry out non-cyclic electron transfer (Section 6.4), use H_2O as electron donor, and are in this respect similar to chloroplasts. Of the remaining groups, the purple bacteria have been the more intensively investigated, and two factors make them suitable for bioenergetic studies. First, mechanical disruption of the cells (for example by extruding a cell suspension through a fine orifice at very high pressure in a French pressure cell) enables the internal membranes to bud off and form isolated closed vesicles called chromatophores (Fig. 1.7). Chromatophores retain the capacity for photosynthetic energy transduction and possess the same orientation as sub-mitochondrial particles. They are the basic units for the study of electron transfer and chemiosmotic energy transduction. The second advantage of the purple bacteria is that the reaction centres (the primary photochemical complexes) can be readily isolated (Section 6.2).

In addition to these bacteria, halobacteria carry out a unique light-dependent energy transduction in which a single protein, bacteriorhodopsin, acts as a light-driven proton pump (Section 6.6).

1.3.5 Reconstituted systems

Reviews Kagawa 1972, Racker 1975

An essential feature of the chemiosmotic theory is that *oxidative phosphoryla-*

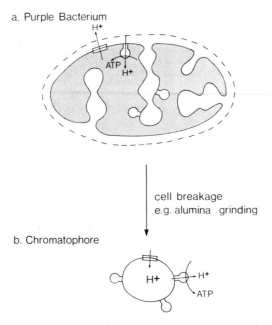

Fig. 1.7 Preparations derived from photosynthetic bacteria.

tion should be functionally and structurally separable into *oxidative, proton-translocating* and *phosphorylating, proton-translocating* complexes, while a similar dissection should be possible in the case of *photosynthetic phosphorylation*. Purification of proton-translocating complexes must be followed by their reincorporation into synthetic, closed membranes in order to observe proton translocation. The value of such "reconstitutions" is twofold. First, it allows for testing aspects of the chemiosmotic theory (Do all energy-transducing complexes pump protons? Does each complex pump protons as an autonomous unit, or only when isolated together with a hypothetical proton pump?). Secondly, it allows the minimum functional unit to be established as a preliminary to investigating the mechanism of the pump.

The complexes are isolated following solubilization in non-denaturing detergents (Section 5.5). They may be reconstituted into a variety of synthetic membrane conformations (Fig. 1.8). Sonication of the complex in the presence of phospholipids produces single-walled micelles or liposomes (Fig. 1.8b). The alternative method of cholate dialysis, in which a suspension of complex and phospholipid in the presence of the bile salt cholate is dialysed in order to slowly lower the detergent concentration, leads to the formation of multilayer liposomes (Fig. 1.8a). Alternatively, liposomes may be fused into planar lipid bilayers to allow direct measurement of the

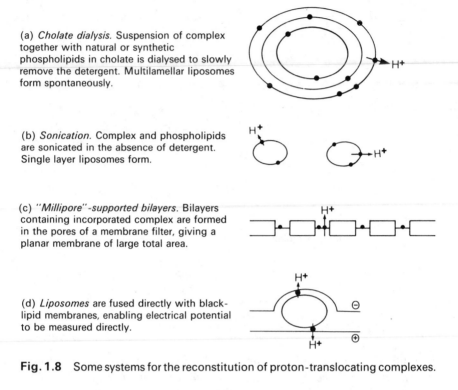

(a) *Cholate dialysis.* Suspension of complex together with natural or synthetic phospholipids in cholate is dialysed to slowly remove the detergent. Multilamellar liposomes form spontaneously.

(b) *Sonication.* Complex and phospholipids are sonicated in the absence of detergent. Single layer liposomes form.

(c) *"Millipore"-supported bilayers.* Bilayers containing incorporated complex are formed in the pores of a membrane filter, giving a planar membrane of large total area.

(d) *Liposomes* are fused directly with black-lipid membranes, enabling electrical potential to be measured directly.

Fig. 1.8   Some systems for the reconstitution of proton-translocating complexes.

electrical potential (Drachev *et al.* 1974; Fig. 1.8d), or a planar bilayer may be formed on a millipore filter (Skulachev 1976, Blok *et al.* 1977; Fig. 1.8c).

The orientation of the incorporated complex can pose problems. Occasionally the high curvature of the membrane imposes a uniform orientation on the incorporated protein, but usually they incorporate randomly. To prevent the complexes working in opposition to each other impermeant substrates must be used, so that only those complexes orientated with their substrate binding sites outwards will be functional.

1.4 BACKGROUND TO THE CHEMIOSMOTIC HYPOTHESIS

Reviews Mitchell 1966, 1968, 1976b, 1979a, b, Greville 1969

This section is intended to provide a brief outline of the relations between the chemiosmotic hypothesis and other hypotheses of energy coupling during the years since its original publication (Mitchell 1961). Such an exercise presents two problems: first, it presupposes a background knowledge of bioenergetics, so that beginners to the field may prefer to omit this section until the concepts become more familiar; secondly, many readers with the necessary background knowledge are likely to disagree with the author's interpretation of this turbulent period. In any case the definitive text remains Greville's classic "worm's-eye scrutiny" of the chemiosmotic hypothesis (Greville 1969).

By the early 1960s the main energy-transducing pathways within the mitochondrion had been established (see Ernster & Lee 1964) except that the central energy-transducing intermediate had not been identified. As a result any hypothesis had to be consistent with a number of basic observations:

(a) The respiratory chain (or "electron-transfer chain") is a linear sequence of electron carriers with three separate regions where redox energy can be conserved in the synthesis of ATP.

(b) The rate of respiration is controlled by the demand for ATP (respiratory control).

(c) Coupling between respiration and ATP synthesis can be disrupted by a group of agents termed "uncouplers" which abolish respiratory control (i.e. stimulate respiration in the absence of ATP synthesis) and allow mitochondria to catalyse a rapid ATP hydrolysis ("uncoupler-stimulated ATPase").

(d) The antibiotic oligomycin (Section 7.2) inhibits both the synthesis and uncoupler-stimulated hydrolysis of ATP.

(e) The energy from respiration can be coupled not only to the synthesis of ATP but also to the accumulation of Ca^{2+} and to the reduction of NAD^+ and $NADP^+$, even though each process is energetically uphill.

(f) These processes can be driven by the hydrolysis of ATP in anaerobic mitochondria, when they can be inhibited by both uncouplers *and* oligomycin.

The pathways of energy transduction (Fig. 1.9a) can be rationalized by postulating a common energy-transducing intermediate, $\sim$ ("squiggle"). The only precedent for such an intermediate comes from glycolysis and the substrate-level phosphorylation catalysed by glyceraldehyde-3-phosphate dehydrogenase and phosphoglycerate kinase (Fig. 1.10). In these reactions, a change in redox state leads to the formation of a phosphate bond with a high negative Gibbs energy of hydrolysis, i.e. a so-called high-energy bond, which could then be transferred to ADP. Adaptation of this "chemical coupling" to the mitochondrion (Slater 1953) required two additional steps,

a. Chemical hypothesis

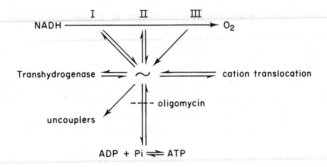

b. Chemiosmotic hypothesis

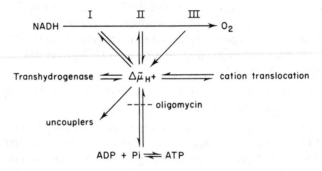

Fig. 1.9 Pathways of mitochondrial energy transduction.

one to allow the three "phosphorylation sites" to generate a common intermediate and one to account for the phosphate-independence of some energy-transducing pathways (Fig. 1.10). This relatively simple formulation was considerably complicated when it was subsequently discovered that mitochondria could accumulate cations such as Ca^{2+} (Section 8.4). To account for this it was necessary to propose a "$\sim$"-driven cation pump (Fig. 1.11c) or, as information about respiration-dependent proton movements grew, a "$\sim$"-driven proton pump (Fig. 1.11b).

It was generally agreed that verification of the chemical hypothesis was contingent on the identification and isolation of one or more of the proposed "high-energy" intermediates. However, despite a number of red herrings, no

a. Substrate-level phosphorylation in glycolysis

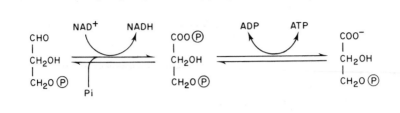

glyceraldehyde-3-phosphate   1,3-diphosphoglycerate   3-phosphoglycerate

i.e.

$$A_{red} + B_{ox} + Pi \rightleftharpoons A_{ox}{\sim}\textcircled{P} + B_{red} \xrightarrow{\text{ADP ATP}} A_{ox}$$

b. Hypothetical chemical coupling in oxidative phosphorylation

$$A_{red} + B_{ox} + I \rightleftharpoons A_{ox}{\sim}I + B_{red}$$

$$A_{ox}{\sim}I + X \rightleftharpoons A_{ox} + X{\sim}I$$

$$X{\sim}I + Pi \rightleftharpoons X{\sim}P + I$$

$$X{\sim}P + ADP \rightleftharpoons X + ATP$$

$$A_{red} + B_{ox} + ADP + Pi \rightleftharpoons A_{ox} + B_{red} + ATP$$

Fig. 1.10 ATP synthesis by chemical coupling.

intermediate could be positively identified, even though the hypothesis predicted that three intermediates would be derivatives of respiratory chain components (such as NAD^+ or a cytochrome) and therefore likely to be amenable to spectroscopic detection.

The chemiosmotic hypothesis did not have its origins in oxidative phosphorylation but had earlier roots in work by Mitchell aimed at understanding the mechanism of "active transport" across cell membranes, in particular how apparently directionless chemical reactions could create directional (vectorial) translocation of a species across the membrane. Mitchell proposed that the structure of transport proteins was such that substrates and products had to approach and leave the catalytic site along defined pathways (Fig. 1.12). In the hypothetical example in Fig. 1.12a, the phosphoryl group "P" undergoes a vectorial transfer across the protein from ATP to "S". Mitchell termed this process "vectorial group translocation". Figure 1.12b shows a

a. Chemiosmotic

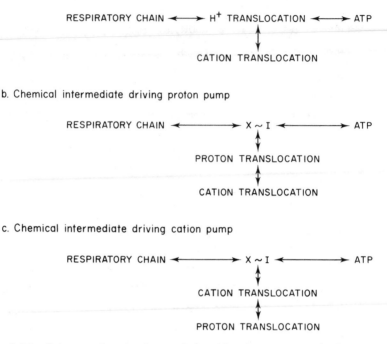

    RESPIRATORY CHAIN ◄───► H⁺ TRANSLOCATION ◄───► ATP

                                          CATION TRANSLOCATION

b. Chemical intermediate driving proton pump

    RESPIRATORY CHAIN ◄───► X ∼ I ◄───► ATP

                               PROTON TRANSLOCATION

                               CATION TRANSLOCATION

c. Chemical intermediate driving cation pump

    RESPIRATORY CHAIN ◄───► X ∼ I ◄───► ATP

                               CATION TRANSLOCATION

                               PROTON TRANSLOCATION

Fig. 1.11 Schemes for the inter-relationships between respiration, proton translocation and cation translocation.

more complex example in which a protein which catalyses the transfer of $2H^+ + 2e^-$ from a reduced substrate AH_2 to B has a structure such that the protons and electrons leave by different routes. This is in fact the basic Mitchell model for proton translocation by the respiratory chain. The electrons travel from donor "A" to acceptor "B" in a "loop" which has two limbs, in the first of which the 2H (or more correctly the $2H^+ + 2e^-$) from AH_2 are bound to a prosthetic group such as FAD. The second limb would be made up by a pure electron-carrying prosthetic group, such as an iron–sulphur centre (Section 5.2). The surplus protons when a "2H"-carrier transfers electrons to an electron carrier would be liberated vectorially on one side of the membrane (Fig. 1.12). Chemiosmosis provided a ready explanation for the energy-dependent transport of ions and metabolites across membranes, transport being either primary (linked directly to metabolism by a group translocation reaction) or secondary (linked indirectly by using the electrochemical gradient generated by the primary event) (see Fig. 1.13).

a. Vectorial transfer of a phosphoryl group `P` from ATP to `S`
(after Mitchell & Moyle 1958)

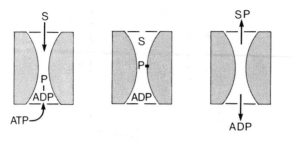

b. Vectorial translocation of protons (after Mitchell 1979 a)

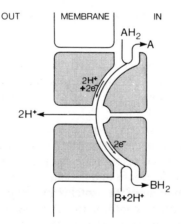

Fig. 1.12 Hypothetical vectorial group translocation models.

Mitchell (1961) emphasized that models for coupling electron flow to proton translocation were not new and could be traced back to the 1940s (Lundegårdh 1945; Fig. 1.14). The novelty of Mitchell's 1961 hypothesis lay in the proposal that the ATP synthetase of mitochondrial and photosynthetic membranes could also translocate protons, hence allowing the $\Delta\tilde{\mu}_{H+}$ generated by electron transfer to drive ATP synthesis (Fig. 1.15).

At the same time as the chemiosmotic hypothesis, Williams (1961) proposed a hypothetical coupling mechanism also involving protons and charge separations, except in this case the protons remained confined to the mem-

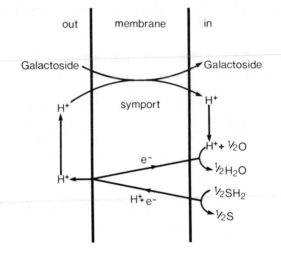

Fig. 1.13 Chemiosmotic scheme for β-galactoside uptake into *E.coli.*

brane and only communicated indirectly with the bulk phases (for reviews see Williams 1976, 1978a). The two authors have been unable to agree over the correct historical priority to be given to the two hypotheses (see Mitchell 1977, Williams 1978b).

The chemiosmotic hypothesis as proposed and developed by Mitchell has remained firmly linked to the concept of vectorial group translocation, in which the proteins of the respiratory chain play largely passive roles as carriers of the prosthetic groups and in defining the vectorial pathways. The idea that the conformations of the proteins may change during oxidation and reduction and that similar conformational changes may occur in the ATP synthetase was first suggested by Boyer (1965). As originally proposed, this "conformational theory" proposed that conformational changes in the respiratory chain during oxidation and reduction were transferred directly to a closely apposed ATP synthetase and that the resultant "strain" induced in the latter provided the energy for ATP synthesis. There was thus no role for the proton electrochemical gradient. Although this direct conformational coupling hypothesis has not stood the test of time (see Boyer *et al.* 1977), the emphasis on protein conformational changes in the molecular mechanism of proton translocation has gained much support as an alternative to vectorial group translocation (Section 5.4).

The great attraction of the chemiosmotic hypothesis to bioenergeticists was that it immediately suggested a large number of experiments by which the proposal could be tested. Three largely independent debates arose as a

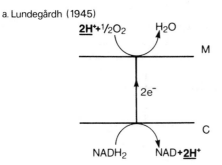

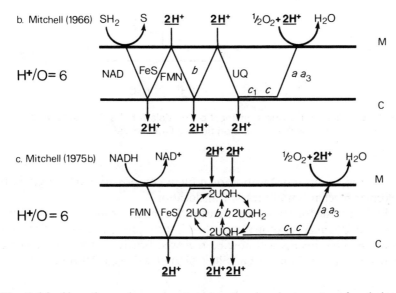

Fig. 1.14 Hypotheses for proton translocation by electron-transfer chains.

consequence. These were first, is the central dogma true and is the proton electrochemical gradient necessary and sufficient for energy transduction? Secondly, is the proton circuit delocalized into the bulk phases as suggested by Mitchell, or are there localized microcircuits of protons between individual respiratory chains and ATP synthetase complexes, i.e. more in accord with the suggestions of Williams? Thirdly, what is the respective importance

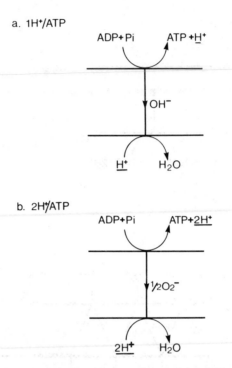

a. 1H⁺/ATP

b. 2H⁺/ATP

Fig. 1.15 Vectorial group translocation models for the ATP synthetase (after Mitchell, as modified by Greville 1969).

of vectorial group translocation or protein conformational changes in the molecular mechanism of proton translocation?

Mitchell made four basic requirements for the experimental verification of the central dogma of proton electrochemical coupling, and some of the more important experiments will now be reviewed briefly. A far more complete treatment is to be found in Greville (1969) or Mitchell (1976b).

(i) *The respiratory and photosynthetic electron transfer chains should translocate protons*

Energy-dependent proton translocation was first reported in "broken" chloroplast preparations (Neumann & Jagendorf 1964). Mitochondrial proton translocation was shown by Mitchell and Moyle (1965, 1967a) by adding a small pulse of O_2 to an anaerobic suspension and following the acidification of the medium (Section 4.3). Low concentrations of uncouplers accelerated the decay of the proton gradient following anoxia. A stoicheiometry of proton extrusion consistent with group translocation (i.e. $2H^+/2e^-$ per loop) was found (but see Section 5.4). During continuous respiration a $\Delta\tilde{\mu}_{H^+}$ of over 200 mV was found (Mitchell & Moyle 1969a; Section 4.2). The direction of proton translocation was reversed in inverted sub-mitochondrial particles (Mitchell & Moyle 1965). Skulachev and colleagues (Skulachev *et al.* 1970) showed that mitochondria could use the proton gradient to accumulate a wide variety of synthetic lipophilic cations, while sub-mitochondrial particles and chromatophores, both with inverted polarity, accumulated the corresponding anions. Since it was unlikely that the mitochondria could have evolved an energy-driven ion pump specific for these synthetic ions, this suggested that a primary cation pump (see Fig. 1.11c) was not possible. Finally, demonstrations that purified respiratory complexes could function as autonomous proton pumps in reconstituted systems (Section 5.5) proved inconsistent with the alternative "chemical hypothesis" of a "~"-driven common proton pump (see Fig. 1.11b).

(ii) *The ATP synthetase should function as a reversible proton-translocating ATPase*

Injection of a small amount of ATP into a suspension of anaerobic mitochondria resulted in an expulsion of protons followed by a slow decay which could be accelerated by uncouplers (Mitchell & Moyle 1968). Purified reconstituted ATP synthetase catalyses a similar proton translocation (Kagawa *et al.* 1973; see Section 7.4). ATP hydrolysis can maintain a $\Delta\tilde{\mu}_{H^+}$ in excess of 200 mV (e.g. Nicholls 1974), while an artificial potential can drive ATP synthesis (Jagendorf & Uribe 1966).

(iii) *Energy-transducing membranes should have a low effective proton conductance*

A low proton permeability can be inferred from the parallel actions of agents which induce proton permeability in synthetic bilayers and at the same time uncouple mitochondria. Proton permeability was measured more directly by Mitchell & Moyle (1967b) by following the rate at which a HCl-induced pH gradient decayed as protons entered the mitochondrial matrix.

(iv) *Energy-transducing membranes should possess specific exchange carriers to permit metabolites to permeate, and osmotic stability to be maintained, in the presence of a high membrane potential*

The decay of ΔpH following a burst of respiration is accelerated by anions such as succinate, Pi and malonate, and by Na^+ (Mitchell & Moyle 1967b), suggesting that anions may be transported with protons and Na^+ in exchange for protons. From the rate of swelling of non-respiring mitochondria in ammonium salts, Chappell and colleagues (Chappell & Crofts 1966, Chappell & Haarhoff 1967, Chappell 1968) obtained evidence for a number of transport systems in which anions are largely transported as uncharged species, either together with protons, or in exchange for other anions (Section 8.3); see also Mitchell & Moyle (1969b).

The second debate on the chemiosmotic hypothesis concerns the extent to which the proton circuit is delocalized over the membrane and into the bulk phases. Local proton circuits have been proposed by a number of workers (e.g. Rottenberg 1975, Kell 1979) on the basis of discrepencies which can sometimes be seen in the relationship between respiration and $\Delta\tilde{\mu}_{H+}$. There are, however, counter-arguments in favour of a proton circuit delocalized over the entire energy-transducing organelle; for instance a single molecule of the proton-translocator gramicidin (Section 2.5) produces detectable uncoupling of an entire chloroplast (Junge & Witt 1968), while in the halophilic bacteria (Section 6.6) the bacteriorhodopsin proton pumps are concentrated in a region of the membrane separated from the ATP synthetase complexes. While the controversy continues, this book will invoke Occam's razor and adopt the simpler, delocalized interpretation.

The third debate, on molecular mechanism, remains open and will continue to do so until we can study the oxidation and reduction of a respiratory chain complex in the same detail that is currently possible for the oxygenation reaction of haemoglobin. Mitchell's group translocation mechanism makes severe demands as regards both stoicheiometry of proton extrusion and sequence of redox carriers (Section 4.3 and 5.4). The current situation (1981) is that there tends to be more support for group translocation among bacterial and photosynthetic workers than there is in the fields of mitochondrial electron transfer or ATP synthetase mechanism.

It must be emphasized again, however, that the central dogma has a firm experimental basis, and the remainder of this book will be based on it.

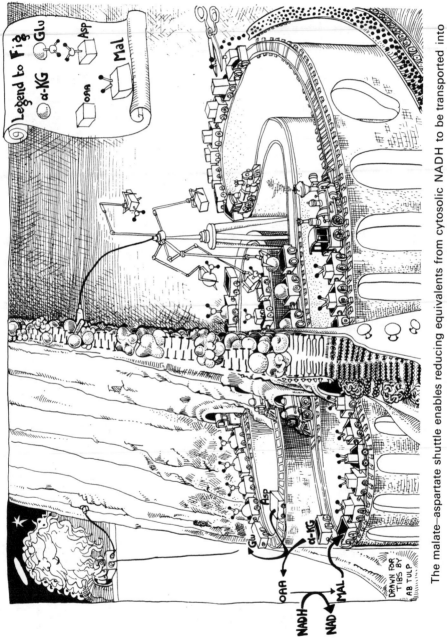

The malate–aspartate shuttle enables reducing equivalents from cytosolic NADH to be transported into the mitochondrial matrix (Fig. 8.2)

2 Ion Transport across Energy-Transducing Membranes

2.1 INTRODUCTION

The advent of the chemiosmotic hypothesis produced a major change in the way in which ion transport processes were regarded. From being a secondary phenomenon, with energy-dependent ion movements being ascribed to side-paths of energy transduction, the study of transport became a central issue in bioenergetics. In particular the ability to correlate the action of ionophores and lipid-soluble ions on energy-transducing membranes with their action in protein-free artificial bilayers was consistent with a primary role of the trans-membrane potential in energy transduction. This chapter describes the basic permeability properties of membranes and the action of ionophores in inducing specific additional pathways of ion permeation.

2.2 THE STRUCTURE OF ENERGY-TRANSDUCING MEM-BRANES

Review DePierre & Ernster 1977

The fluid-mosaic model of membrane structure (Singer & Nicolson 1972) is adequate to explain the properties of energy-transducing membranes (Fig. 2.1). In this model the majority of the phospholipid is arranged as a bilayer with polar head-groups at the two aqueous interfaces. The membrane proteins may be either peripheral (extrinsic) or integral (intrinsic), depending on the depth to which they are buried within the hydrophobic core of the bilayer. Some integral proteins may span the membrane, and allow protein-catalysed transport across the membrane. The distinction between protein-catalysed transport and transport across bilayer regions of the membrane will be emphasized in this chapter.

While the fluid-mosaic model is usually represented with protein "icebergs" floating in a sea of lipid, the high proportion of protein in energy-transducing membranes (in the case of the mitochondrial inner membrane 50% of the membrane is integral protein, 25% peripheral protein and 25%

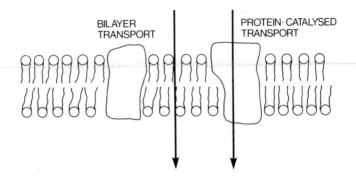

Fig. 2.1   The fluid-mosaic model.
A consequence of the fluid-mosaic model is that transport can occur either across lipid bilayer regions of the membrane, or through proteins. The characteristics of these two modes are quite distinct.

lipid) results in a relatively close packing of the proteins, less than 60% of the membrane being a bilayer. Energy-transducing membranes also tend to have distinctive lipid compositions: 10% of the mitochondrial inner membrane lipid is cardiolipin; in the case of the chloroplast thylakoid membrane only 10% of the lipid is phospholipid, the remainder being galactolipids (40%) sulpholipids (4%) and photosynthetic pigments (40%). Despite this heterogeneity of lipid composition, the native and ionophore-induced permeability properties of the bilayer regions of the different membranes are sufficiently similar to justify extrapolations between energy-transducing membranes and artificial bilayer preparations. However, protein-catalysed transport can be unique, not only to a given organelle but also to an individual tissue. For example the inner membrane of rat liver mitochondria possesses protein-catalysed transport properties which are absent in mitochondria from rat heart (Section 8.3).

2.3 PATHWAYS OF ION TRANSPORT

For an ion to be transported across a membrane both a pathway and a driving force are required. Driving forces can be concentration gradients, electrical potentials, metabolic energy, or combinations of these. They will be discussed in Chapter 3; this chapter will deal with the natural and induced pathways which exist in energy-transducing membranes.

There are a number of questions which can be asked about any transport process:

(a) *Does the ion cross the membrane on its own, or is its transport directly coupled to the movement of another ion (Fig. 2.2)?*

A transport process involving a single ion is termed a *uniport*. Examples of uniports include the uptake pathway for Ca^{2+} across the inner mitochondrial membrane (Section 8.4) and the proton permeability induced in bilayers by the addition of proton translocators such as dinitrophenol (Section 2.5). A transport process involving the obligatory coupling of two or more ions in parallel is termed *symport* or *co-transport*. A number of cases of H^+/metabolite symport occur across the bacterial plasma membrane (Section 8.5). The equivalent tightly coupled process where the transport of one ion is linked to the transport of another species in the opposite direction is termed *antiport* or *exchange-diffusion*, examples include the Na^+/H^+ antiport activity which is present in the inner mitochondrial membrane (Section 2.6) and the K^+/H^+ antiport catalysed by the ionophore nigericin in bilayers (Section 2.5). If one of the ions involved in a nominal symport or antiport mechanism is a H^+ or OH^-, it is usually impossible to distinguish between the symport of a species with a H^+ and the antiport of the species with a OH^-. For example the mitochondrial phosphate carrier (Section 7.5) may be variously represented as a Pi^-/OH^- antiport or a H/Pi symport.

(b) *Do the directly coupled ion movements result in the transfer of charge across the membrane (Fig. 2.2)?*

Electroneutral transport involves no net charge transfer across the membrane. Transport may be electroneutral either because an uncharged species is transported by a uniport or as the result of the symport of a cation and an anion or the antiport of two ions of equal charge, an example of the last being the K^+/H^+ antiport catalysed by nigericin. *Electrical* transport is frequently termed either *electrogenic* ("creating a potential") or *electrophoretic* ("moving in response to a pre-existing potential"). As these terms can refer to the same pathway observed under different conditions the overall term "electrical" will be used here.

It is important to distinguish between electrical balance at the molecular level, as discussed here, and the overall electroneutrality of the total ion movements across a given membrane. The latter follows from the impossibility of separating more than minute quantities of positive and negative charge across a membrane without building up a large membrane potential. Thus the separation of 1 nmol of charge across the inner membranes of 1 mg of mitochondria results in the build-up of more than 200 mV of potential (Section 4.2). However, this does not preclude the occurrence of individual electrical events at the molecular level as long as these compensate each other

A (i) Uniport

e.g.

(ii) Symport

e.g.

(iii) Antiport

e.g.

B (i) Electroneutral

e.g.

or

(ii) Electrical

e.g.

or

C (i) Not coupled directly to metabolism
                                      e.g. any of the examples in A and B
   (ii) Coupled to metabolism

e.g.

or

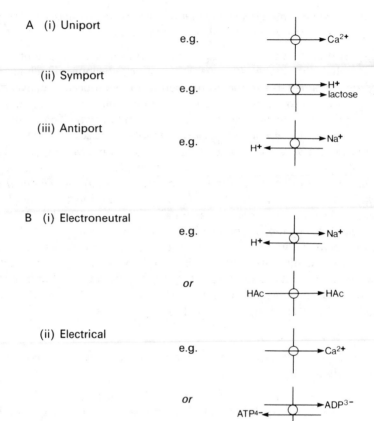

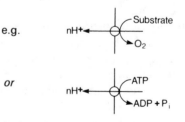

Fig. 2.2 Further classifications of ion transport.

(Section 2.7). In addition it is necessary to appreciate that the effect on an energy-transducing membrane of a tightly coupled electroneutral antiporter is not the same as that caused by the addition of two electrical uniporters for the same ions.

(c) *Is transport directly coupled to metabolism (Fig. 2.2)?*

A tight coupling of transport to metabolism occurs in the ion pumps which are central to chemiosmotic energy transduction. The term "active transport" is sometimes applied in this context, although it is important to restrict the term to examples where such inter-conversions occur, and not merely to include all cases where ions are concentrated across a membrane. Ions can be accumulated across a membrane without invoking an ion pump either if there is a membrane potential or if transport is coupled by symport or antiport to the "downhill" movement of a second ion. For example, while Ca^{2+} is accumulated within the sarcoplasmic reticulum by an ion pump (the Ca^{2+}–ATPase), the same ion is accumulated across the mitochondrial inner membrane as a consequence of the membrane potential (Section 8.4). Only the former case could be described as "active", mitochondrial Ca^{2+} accumulation occurring down the electrochemical gradient.

Group translocation (Section 1.4) represents a specific proposal for the molecular mechanism of coupling of metabolism to ion transport.

(d) *Does transport occur across bilayer or protein regions of the membrane (Fig. 2.1)?*

This final classification stems directly from the fluid-mosaic model of the membrane. The distinction is crucial, since bilayer transport can be generalized for different membranes, whereas protein-catalysed transport must be considered for specific membranes. The characteristics of these two forms of transport are sufficiently distinct to merit separate discussion.

2.4 THE NATURAL PERMEABILITY PROPERTIES OF BILAYER REGIONS

The hydrophobic core possessed by lipid bilayers creates an effective barrier to the passage of charged species. With a few important exceptions (Section 2.5) bilayers are therefore impermeable to cations and anions. This impermeability extends to the proton, and this property is vital for energy transduction to avoid short-circuiting the proton circuit (Section 4.5). Not only does the bilayer have a high electrical resistance, but it can also withstand very high electrical fields. An energy-transducing membrane with a membrane potential of 200 mV has an electrical field in excess of 300 000 V cm^{-1} across its hydrophobic core.

A variety of uncharged species can cross bilayers. H_2O, O_2 and CO_2 are all highly permeable, as are the uncharged forms of a number of low molecular weight acids and bases, such as ammonia and acetic acid. These last permeabilities provide a useful tool for the investigation of pH gradients across membranes (Section 3.5).

2.5 IONOPHORE-INDUCED PERMEABILITY PROPERTIES OF BILAYER REGIONS

Reviews Henderson 1971, Pressman 1976, Gomez–Poyou & Gomez–Lojero 1977, Ovchinnikov 1979

The high activation energy required to insert an ion into a hydrophobic region is the reason for the extremely low ion permeability of bilayer regions. It follows that if the charge can be delocalized and shielded from the bilayer, the ion permeability might be expected to increase. This is accomplished by a variety of antibiotics synthesized by some micro-organisms, as well as by some synthetic compounds. These are known collectively as ionophores. Ionophores are typically compounds with a molecular weight of 500–2000 and possess a hydrophobic exterior, to be lipid soluble, together with a hydrophilic interior to bind the ion. There is little evidence that ionophores are natural constituents of energy-transducing membranes, but as tools for investigation they are invaluable.

There are two mechanisms by which ionophores can function, as mobile carriers or as channel formers (Fig. 2.3). Mobile carriers diffuse within the membrane, and can typically catalyse the transport of about 1000 ions per second across the membrane. They can show an extremely high discrimination between different ions, can work across thick synthetic membranes, and are affected by the fluidity of the membrane. In contrast, channel-forming ionophores show a poor discrimination between ions, but can be very active, catalysing the transport of up to 10^7 ions per channel per second.

2.5.1 Carriers of charge but not protons

Valinomycin

Valinomycin (Fig. 2.4) is a mobile carrier ionophore which catalyses the electrical uniport of Cs^+, Rb^+, K^+ or NH_4^+. The ability to transport Na^+ is at least 10^4 less than for K^+. Valinomycin is a natural antibiotic from *Streptomyces* and is a depsipeptide, i.e. it consists of alternating hydroxy-

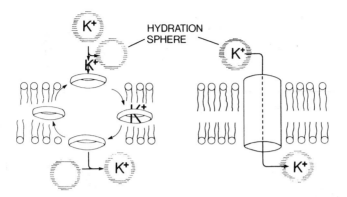

Fig. 2.3 (*Left*) Mobile-carrier (e.g. valinomycin) and (*right*) channel-forming ionophores.

and amino acids. The ions lose their water of hydration when they bind to the ionophore (Fig. 2.3). Na^+ cannot be transported because the unhydrated Na^+ ion is too small to interact efficiently with the inward-facing carbonyls of valinomycin, with the result that the complexation energy does not balance that required for the loss of the water of hydration. Because valinomycin is uncharged and contains no ionizable groups, it acquires the charge of the complexed ion. Both the uncomplexed and complexed forms of valinomycin are able to diffuse across the membrane, therefore a catalytic amount of ionophore can induce the bulk transport of cations. It is effective in concentrations as low as 10^{-9} M in mitochondria, chloroplasts, synthetic bilayers and to a more limited extent in bacteria.

Energy-transducing membranes generally lack a native electrical K^+ permeability, the value of the ionophore lies in the ability to induce such a permeability in order to estimate (Section 4.2) or abolish (Section 4.3) membrane potentials, or to investigate anion transport (Section 2.7).

Other ionophores catalysing K^+ uniport include the enniatins and the nactins (nonactin, monactin, dinactin, etc., so-called from the number of ethyl groups in the structure). However, these ionophores do not have such a spectacular selectivity for K^+ over Na^+ as valinomycin.

Gramicidin

Gramicidin is a channel-forming ionophore (Fig. 2.4) which forms transient conducting dimers in the bilayer. Its properties are typical of channel-forming ionophores, with a poor selectivity between protons, monovalent cations and NH_4^+, the ions permeating in their hydrated forms. The capacity to conduct ions is limited only by diffusion, with the result that one channel can conduct up to 10^7 ions s^{-1}.

(a) Valinomycin

Cation specificity: $Rb^+ > K^+ \gg Na^+ > Li^+$

(b) Nigericin

Cation specificity: $K^+ > Rb^+ > Na^+ > Li^+$

(c) Gramicidin

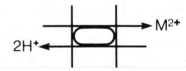

Cation specificity: $H^+ > Rb^+ > K^+ > Na^+ > Li^+$

(d) A23187

Cation specificity: $Ca^{2+} > Mg^{2+}$

Fig. 2.4   Some ionophore-catalysed transport processes.

2.5.2 Carriers of protons but not charge

Nigericin

Nigericin is a linear molecule with heterocyclic oxygen-containing rings together with a hydroxyl group. In the membrane the molecules cyclize to form a structure similar to that of valinomycin, with the oxygen atoms forming a hydrophobic interior. Unlike valinomycin, nigericin loses a proton when it binds a cation, forming a neutral complex which can then diffuse across the membrane as a mobile carrier. Nigericin is also mobile in its protonated non-complexed form, with the result that the ionophore can catalyse the overall electroneutral exchange of K^+ for H^+ (Fig. 2.4). Other ionophores which catalyse a similar electroneutral exchange include X-537A, monensin, and dianemycin. The latter two show a slight preference for Na^+ over K^+, while X-537A will complex virtually every cation, including organic amines.

Nigericin has been employed to study anion transport (Section 2.7) and to modify the pH gradient across energy-transducing membranes. It is often stated that nigericin abolishes ΔpH across a membrane; in fact the ionophore equalizes the K^+ and H^+ gradients, the final ion gradients depending on the experimental conditions.

A23187

A23187 is a carboxylic ionophore with a high specificity for divalent rather than monovalent cations. It catalyses the electroneutral exchange of Ca^{2+} or Mg^{2+} for two H^+ without disturbing monovalent ion gradients.

2.5.3 Carriers of protons and charge

Proton translocators ("uncouplers")

Proton translocators have dissociable protons and are permeable across bilayers either as protonated acids or as the conjugate base (Fig. 2.5). This is possible because these ionophores possess extensive π-orbital systems which so delocalize the charge of the anionic form that lipid solubility is retained. By shuttling across the membrane they can catalyse the net electrical uniport of protons and increase the proton conductance of the membrane. In so doing the proton circuit is short-circuited (Section 4.5), allowing the generator of $\Delta\tilde{\mu}_{H+}$ to be uncoupled from the ATP synthetase. The un-coupling action of certain compounds was described long before the formula-tion of the chemiosmotic theory; the demonstration that the majority of these compounds act by increasing the proton conductance of synthetic bilayers was an important piece of evidence in favour of the theory and against chemical intermediate theories which had assigned a specific role to uncouplers in hydrolysing hypothetical high-energy intermediates.

An indirect proton translocation can be induced in membranes by the combination of a uniport for an ion together with an electroneutral antiport of the same ion in exchange for a proton. For example the combination of valinomycin and nigericin induces a net uniport for H^+, while K^+ cycles around the membrane.

e.g. carbonyl cyanide-*p*-trifluoromethoxyphenylhydrazone (FCCP)

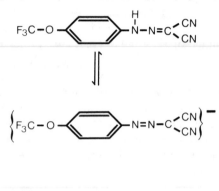

Fig. 2.5   Proton translocators catalyse proton uniport across the membrane. Proton translocators (uncouplers) are lipophilic weak acids permeable across lipid bilayers in either the protonated or deprotonated forms. If a proton electrochemical potential (Sec. 3.4) exists across the membrane the proton translocator will cycle catalytically in an attempt to collapse the potential; $FCCP^-$ will be driven to the positive face of the membrane by the membrane potential, Sec. 3.4, while FCCPH will be driven towards the alkaline compartment due to the pH gradient, Sec. 3.5. Uncouplers such as FCCP function in a wide variety of membranes at concentrations from 10^{-9} to 10^{-5}M.

2.5.4 Lipophilic cations and anions

The ability of π-orbital systems to delocalize charge and enhance lipid solubility has been exploited in the synthesis of a number of cations and anions which are capable of being transported across bilayer membranes even though they carry charge (Fig. 2.6). These ions are not strictly ionophores, since they do not act catalytically, but are instead accumulated in response to $\Delta\Psi$ (Section 4.4). Lipophilic cations and anions have been of great value

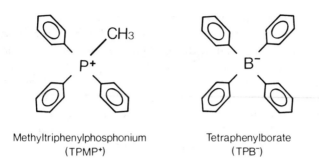

Methyltriphenylphosphonium
(TPMP⁺)

Tetraphenylborate
(TPB⁻)

Fig. 2.6 Examples of synthetic lipophilic ions.

historically, in demonstrations of their energy-dependent accumulation in mitochondria and inverted sub-mitochondrial particles respectively (Section 1.4). These experiments eliminated the possibility of specific cation pumps driven by high-energy intermediates (Skulachev 1971). Subsequently the cations have been employed for the estimation of $\Delta\Psi$ (Section 4.2).

2.6 PROTEIN-CATALYSED TRANSPORT

Review Wilbrant 1974

The characteristics of protein-catalysed transport across energy-transducing membranes are usually sufficiently distinct from those of bilayer-dependent transport, whether in the absence or presence of ionophores, to make the correct assignment straightforward. Transport proteins share the features of other enzymes: they can display stereospecificity, can frequently be inhibited specifically, and are genetically determined. This last feature means that it is not possible to make the same kinds of generalizations as for bilayer transport. For example, if FCCP (Fig. 2.5) induces proton permeability in mitochondria, it can generally be assumed that the effect will be the same on chloroplasts, bacteria and synthetic bilayers. In contrast, a transport protein may not only be specific to a given organelle but may be restricted to the organelle from one tissue. For example the citrate carrier is present in liver mitochondria, where it is involved in the export of intermediates for fatty acid synthesis (Section 8.3), but is absent from heart mitochondria. It is sometimes stated that saturation kinetics are characteristic of protein-mediated transport. Although this may be true on occasions, the kinetics of any transport process is so complex, particularly if a membrane potential is applied, that kinetics must be interpreted with care.

The strongest evidence for the involvement of a protein is the existence of specific inhibitors. For example, pyruvate was for many years considered to permeate into mitochondria through bilayers, which is feasible as it is a monocarboxylic weak acid. However, it was found that cyanohydroxy-cinnamate (Section 8.3) was a specific transport inhibitor, and this provided the first firm evidence for a carrier (see LaNoue & Schoolworth 1979).

Transport proteins have been studied by many approaches, and this has led to a plethora of names, including carriers, permeases, porters and translocases, all of which are synonyms for transport protein.

2.7 BULK SOLUTE MOVEMENTS ACROSS ENERGY-TRANSDUCING MEMBRANES

Suspensions of mitochondria are turbid and scatter light. The light scattered is largely a function of the difference in refractive index between the matrix contents and the medium, and any process which decreases this difference will decrease the scattered light. Paradoxically, an increase in the matrix volume due to the influx of a permeable solute results in a decrease in the light scattered as the matrix refractive index approaches that of the medium. This provides a very simple qualitative method for the study of solute fluxes across the mitochondrial inner membrane. Mitochondria are suited for this technique since their matrices can undergo a large increase in volume without bursting the inner membrane, as the inner membrane simply unfolds its cristae. "Swelling" can proceed sufficiently to rupture the outer membrane and release adenylate kinase, which is located in the inter-membrane space. Light scattering can be followed either from the decrease in transmitted light in a normal spectrophotometer, or more sensitively by using the 90° geometry of a fluorimeter to measure the scattered light directly.

To observe osmotic swelling of mitochondria in ionic media, two criteria must be satisfied: first, both the cation and anion of the major osmotic component of the medium must be permeable; secondly the requirement for overall charge balance across the membrane must be respected. The simplest case to consider is that of mitochondria where neither the respiratory chain nor the ATP synthetase are functional. In the example shown in Fig. 2.7 the permeability of rat liver mitochondria to Cl^- and CNS^- is investigated. Mitochondria suspended in either 120 mM KCl or 120 mM KCNS undergo little swelling. However, this does not demonstrate that the inner membrane is impermeable to these anions, as the membrane is poorly permeable to K^+. To overcome this, an electrical uniport for K^+ can be induced by the addition of valinomycin (Section 2.5), which as an ionophore can be confidently pre-dicted to induce the same permeability in the lipid bilayer region of this mem-brane as it does in synthetic membranes. Rapid swelling is now observed in

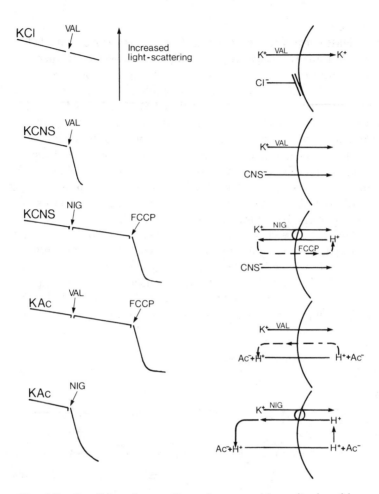

Fig. 2.7 Conditions for swelling of non-respiring mitochondria.

KCNS, but slow swelling in KCl. It might therefore be concluded that the inner membane is permeable to CNS$^-$ but not to Cl$^-$. As luck would have it, this conclusion turns out to be correct although premature, as the problems raised by charge balance have yet to be considered.

To illustrate this requirement, swelling in K-acetate is also shown in Fig. 2.7. In order to ensure that there is no rate limitation due to K$^+$ impermeability, ionophores (valinomycin or nigericin) are present. However, it is clear that the rate of swelling depends on the nature of the permeability induced by the ionophore: nigericin is only effective with K-acetate, while valinomycin only allows rapid swelling to occur in KCNS. The reason is the need for charge balance. K^+ entry catalysed by valinomycin is electrical, while acetate permeates the bilayer as the neutral protonated acid. Therefore a large $\Delta\Psi$ (positive inside) rapidly builds up preventing further K^+ entry. The permeation of acetic acid also ceases, as dissociation of the co-transported proton within the matrix builds up a pH gradient (acid in the matrix) which opposes further acetic acid entry (Section 3.5). These problems are not encountered with nigericin, first because cation and anion entry are both now electroneutral and secondly because the proton entering with acetic acid is re-exported by the ionophore in exchange for K^+. For similar reasons, mitochondria swell in KCNS plus valinomycin, but not in KCNS plus nigericin. It is therefore possible to use swelling, not only to determine if a species is permeable but also to determine the mode of entry. For example, Ca^{2+} entry into non-respiring mitochondria has been shown to be consistent with a Ca^{2+} uniport by such a technique (Selwyn *et al.* 1970; see Section 8.4).

Swelling therefore requires that both ions enter by the same mode, be that electrical or electroneutral. There is, however, a way to induce swelling when this requirement is not met—to induce an electrical proton uniport by the addition of a proton translocator such as FCCP (Fig. 2.7). The fact that classical uncouplers of oxidative phosphorylation (Section 1.4) act as pre-

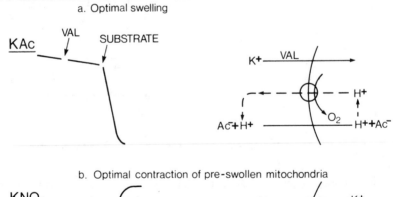

a. Optimal swelling

b. Optimal contraction of pre-swollen mitochondria

Fig. 2.8   Respective conditions for optimal swelling and optimal contraction of respiring mitochondria.

dicted by the chemiosmotic theory in non-respiring mitochondria (where there can be no hypothetical high-energy intermediates) was important evidence in favour of the theory (Section 1.4). Brown adipose tissue mitochondria possess a native proton short-circuit in their inner membrane (Section 4.5) which was first revealed by this technique.

Matrix volume changes occurring in respiring mitochondria have to take account of the contribution of the protons pumped across the membrane by the respiratory chain. Respiration-dependent swelling occurs in the presence of an electrically permeant cation (which is accumulated due to the membrane potential) and an electroneutrally permeant weak acid (accumulated due to ΔpH; Section 3.5), as shown in Fig. 2.8. Conversely, a rapid contraction of pre-swollen mitochondria occurs on initiation of respiration when the matrix contains an electroneutrally permeant cation (expelled by ΔpH) and an electrically permeant anion (expelled by $\Delta\Psi$).

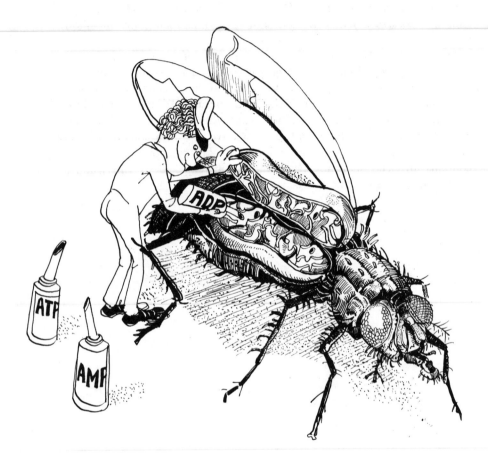

AB TULP

3 Quantitative Bioenergetics: the Measurement of Driving Forces

3.1 INTRODUCTION

This chapter is intended to provide an introduction to that part of thermodynamics which is of relevance to bioenergetics. This is the most frequently misunderstood aspect of bioenergetics, since most biochemists (like the author) have little background in physical chemistry. The topics have been limited to those of specific bioenergetic relevance, but at the same time I have attempted to de-mythologize some of the more important relationships by deriving them from what I hope are commonsense origins. The reader is strongly advised to follow through the derivations, if only to free himself of the idea that ATP is a "high-energy" compound. However I have also indicated the most essential sections for rapid reference by enclosing them in boxes.

In thermodynamics three types of system are studied: isolated (or adiabatic) systems are completely autonomous, exchanging neither material nor energy with their surroundings; closed systems are materially self-contained, but are capable of exchanging energy across their boundaries; open systems exchange both energy and material with their environment. The complexity of the thermodynamic treatment of these systems increases as their isolation decreases. Classical equilibrium thermodynamics cannot be applied precisely to open systems because the flow of matter across their boundaries precludes the establishment of a true equilibrium within; as all bioenergetic systems are open, continually exchanging substrates and end-products with their environment, it is important to appreciate that there are restrictions in the application of equilibrium thermodynamics to the field.

Classical thermodynamics remains of use, however, in the prediction of the conditions required for equilibrium in a complex energy transduction, and by extension in predicting the direction in which energy flows when the reaction is displaced from equilibrium. While no thermodynamic treatment can prove the existence of a given mechanism, equilibrium thermodynamics can readily disprove any proposed mechanism which disobeys its laws.

41

Finally, under near-equilibrium conditions, equilibrium parameters can be inserted into the equations of irreversible thermodynamics (which deals with open systems) to give information on the rate of energy flow.

In an isolated system the driving force for a reaction is entropy, and only if a process increases the entropy of the system can it occur spontaneously. In a closed system the entropic driving force is still present, but one must consider the net entropy change of the system and its surroundings together in order to predict spontaneity. Because entropy changes in the surroundings are rarely measurable, it is necessary to calculate this net entropy change purely in terms of measurable parameters within the system, namely the entropy change within the system, and the flow of enthalpy (heat) across the boundaries of the system (Fig. 3.1). In both isolated and closed systems the Gibbs energy change, ΔG is the quantitative measure of the net driving force (at constant temperature and pressure), and a process which results in a decrease in Gibbs energy ($\Delta G < 0$) is one which causes a net increase in the entropy of the system plus surroundings and can therefore occur spontaneously if a mechanism is available.

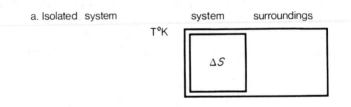

$$\Delta G = -T\Delta S$$

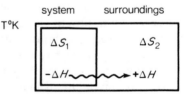

if $\Delta S = \Delta S_1 - \Delta S_2$
$\Delta G = -T\Delta S = T\Delta S_2 - T\Delta S_1$
but $\Delta S_2 = \Delta H/T$
$\Delta G = \Delta H - T\Delta S_1$

Fig. 3.1   Gibbs energy changes in isolated and closed systems.

The Gibbs energy change (also termed the free energy change) occurs in bioenergetics in four different guises; indeed the subject might well be defined as the study of the mechanisms by which the different manifestations of Gibbs energy are interconverted. (1) Gibbs energy changes themselves are used in the description of substrate reactions feeding into the respiratory chain and of the ATP which is ultimately synthesized. (2) The oxido-reduction reactions occurring in the electron transfer pathways in respiration and photosynthesis are usually quantified not in terms of Gibbs energy changes but in terms of closely derived redox potential changes. (3) The available energy in the proton gradient is quantitated by a further variant of the Gibbs energy change, namely the proton electrochemical potential. (4) In photosynthetic systems the Gibbs energy available from the absorption of quanta of light can be compared directly with the other Gibbs energy functions. It should be emphasized that these different conventions merely reflect the diverse historical background of the elements which are brought together in chemiosmotic energy transduction.

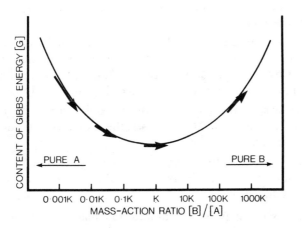

Fig. 3.2 Gibbs energy of a reaction as a function of the displacement from equilibrium.
A closed system contains components A and B at concentrations [A] and [B]. A and B can be interconverted by a reaction A⇌B. This reaction is at equilibrium when the mass–action ratio [B]/[A]$=K$. The curve shows qualitatively how the Gibbs energy of the system (G) varies when the total [A]+[B] is held constant, but the mass–action ratio is varied away from equilibrium. The arrows represent schematically the Gibbs energy *change* (ΔG) for a small conversion of A to B occuring at different mass–action ratios.

3.2 GIBBS ENERGY

Consider a simple reaction A⇌B, possessing a characteristic equilibrium constant K. If it were possible to determine the absolute value of the Gibbs energy (G) as a function of the extent of reaction, the curve would be a parabola of the form shown in Fig. 3.2. The curve shows the following features:

(a) The Gibbs energy content (G) is at a minimum when the reaction is at equilibrium. Thus any change in the observed mass–action ratio [B]/[A] away from the equilibrium ratio would require an increase in the Gibbs energy content of the system and so cannot occur spontaneously.

(b) The slope of the curve is zero at equilibrium. This means that a small conversion of A to B which occurs without significantly changing the mass–action ratio would cause no change in the Gibbs energy content, i.e. the slope ΔG (in units of kJ mole^{-1}) is zero at equilibrium.

(c) When the reaction A→B has not yet proceeded as far as equilibrium, a small conversion of A to B results in a decrease in G, i.e. the slope ΔG is negative implying that such an interconversion can occur spontaneously, provided that a mechanism exists.

(d) The slope of the curve decreases as equilibrium is approached. This implies that ΔG|decreases the closer the reaction is to equilibrium.

(e) For the reaction to proceed beyond the equilibrium point would require an input of Gibbs energy, this therefore cannot occur spontaneously.

The discussion may be generalized and placed on a quantitative footing by considering the reaction:

$$aA + bB \rightleftharpoons cC + dD \qquad \text{(Eq. 3.1)}$$

The equilibrium constant for the reaction is defined as follows:

$$K = \frac{[C]_{eq}^{c} \cdot [D]_{eq}^{d}}{[A]_{eq}^{a} \cdot [B]_{eq}^{b}} \text{Molar}^{(c+d-a-b)} \qquad \text{(Eq. 3.2)}$$

the equilibrium concentration of each component being inserted into the equation to obtain an equilibrium mass–action ratio.

If the observed mass–action ratio when the reaction is held away from equilibrium is defined by

$$\Gamma = \frac{[C]_{obs}^{c} \cdot [D]_{obs}^{d}}{[A]_{obs}^{a} \cdot [B]_{obs}^{b}} \; \text{Molar}^{(c+d-a-b)} \qquad \text{(Eq. 3.3)}$$

then the Gibbs energy change when a moles of A and b moles of B are converted to C and D without significantly changing Γ is given by:

$$\Delta G = -2 \cdot 3 RT \log_{10} K/\Gamma \qquad \text{(Eq. 3.4)}$$

(where the factor 2·3 comes from the conversion from natural logarithms, R is the gas constant, and T is the absolute temperature) ΔG therefore has a value which is a function of the displacement from equilibrium, the factor $2 \cdot 3RT$ meaning that at 25°C a reaction which is maintained one order of magnitude away from equilibrium possesses a ΔG of 5·7 kJ mole^{-1}, the Gibbs energy function being negative if the observed mass–action ratio is less than the equilibrium constant, and positive if the mass–action ratio is greater.

This relationship may be illustrated by reference to the hydrolysis of ATP to ADP and Pi. At pH 7·0, and in the presence of 10mM Mg^{2+}, this reaction has an apparent equilibrium constant K' of about 10^5 (Rosing and Slater, 1972):

$$K' = \frac{[\Sigma ADP][\Sigma Pi]}{[\Sigma ATP]} = 10^5 \qquad \text{(Eq. 3.5)}$$

where each concentration represents the total sum of the concentrations of the different ionized species of each component, including that complexed to Mg^{2+} (see below). As equilibrium is attained when Γ is 10^5, in the presence of 10 mM Pi and 10 mM ADP, which are typical figures for the cytoplasm, the equilibrium concentration of ATP would be only 10^{-9} M.

The variations of ΔG with the displacement of the mass-action ratio from equilibrium are shown in Table 3.1. Mitochondria are able to maintain a mass-action ratio in the incubation medium which is as low as 10^{-5} M, ten orders of magnitude away from equilibrium (Slater et al. 1973). Under these conditions the incubation might contain 10 mM Pi, 10 mM ATP and only 10^{-5} M ADP. Each additional mol of ATP synthesized by the mitochondria while maintaining these concentrations would, from Table 3.1, require an input of 57 kJ mol^{-1} of Gibbs energy. Note that the Gibbs energy change for

ATP synthesis is obtained from the corresponding value for ATP hydrolysis by simply changing the sign.

Table 3.1   The Gibbs energy change for the hydrolysis of ATP to ADP+Pi as a function of the displacement from equilibrium[a]

If $K'_{(pH\ 7,\ [Mg^{2+}]\ 10\ mM)} = 1.0 \times 10^5$

Observed mass–action ratio (Γ)	K'/Γ	ΔG (kJ mol^{-1})	[ATP]/[ADP] when [Pi]=10 mM
10^5	1	0	10^{-7}
10^3	10^2	-11.4	10^{-5}
10	10^4	-22.8	10^{-3}
1	10^5	-28.5	10^{-2}
0.1	10^6	-34.2	10^{-1}
10^{-3}	10^8	-45.6	10
10^{-5}	10^{10}	-57	10^3

[a] For exact values see Rosing & Slater 1972

A special case of the general equation for ΔG (Eq. 3.4) occurs when the observed mass–action ratio of a reaction is unity. These conditions define the standard Gibbs energy change ΔG^0, and the equation reduces to:

$$\Delta G^0 = -2.3RT \log_{10} K \qquad \text{(Eq. 3.6)}$$

By substituting this relationship into Eq. 3.4. the equilibrium constant may be eliminated:

$$\Delta G = \Delta G^{0|} + 2.3RT \log_{10} \Gamma \qquad \text{(Eq. 3.7)}$$

This is most commonly encountered form of the Gibbs energy equation. Note that K and Γ are dimensionless by this treatment.

To avoid confusion or ambiguity in the derivation of equilibrium constants, and hence Gibbs energy changes, a number of conventions have been adopted. Those most relevant to bioenergetics are the following:

(a) True thermodynamic equilibrium constants are defined in terms of the chemical activities rather than the concentrations of the reactants and products. Generally in biochemical systems it is not possible to determine the activities of all the components, and so equilibrium constants are calculated from concentrations. This introduces no error as long as the observed mass–action ratio and the equilibrium constants are calculated under comparable conditions.

(b) When water appears as either a reactant or product, its concentration is taken as unity, rather than 55 M. This means that the water term can be omitted from the equilibrium and observed mass–action ratio equations (see e.g. Eq. 3.5).

(c) If one or more of the reactants or products are ionizable, or can chelate a cation, there is an ambiguity as to whether the equilibrium constant should be calculated from the total sum of the concentrations of the different forms of a compound, or just from the concentration of that form which is believed to participate in the reaction. The hydrolysis of ATP to ADP and Pi is a particularly complicated case: not only are all the reactants and products partially ionized at physiological pH, but also Mg^{2+}, if present, chelates ATP and ADP with different affinities (see Chappell 1977). Thus ATP can exist at pH 7 in the following forms:

$$[\Sigma ATP] = [ATP^{4-}] + [ATP^{3-}] + [ATPMg^{2-}] + [ATPMg^{-}] \quad \text{(Eq. 3.8)}$$

If it were known that the true reaction was:

$$ATPMg^{2-} + H_2O = ADPMg^{-} + Pi^{2-} + H^{+} \quad \text{(Eq. 3.9)}$$

then the true equilibrium constant would be:

$$K = \frac{[ADPMg^{-}][Pi^{2-}][H^{+}]}{[ATPMg^{2-}]} \quad \text{(Eq. 3.10)}$$

This equilibrium constant would be independent of pH or Mg^{2+}, as these factors are allowed for in the equation. However, the reacting species are not known unambiguously, and even if they were, their concentrations would be difficult to assay, as enzymatic or chemical assay determines the total concentration of each compound (e.g. ΣATP).

In practice, therefore, an apparent equilibrium constant, K' is employed, calculated from the total concentrations of each reactant and product, ignoring any effects of ionization or chelation (see Eq. 3.5). In particular any protons which are involved are omitted from the equation (see Eq. 3.5).

The most important limitation is that K' is not a constant for a given reaction at constant temperature, but depends on all those factors which are omitted from the equation, e.g. pH and cation concentration. Thus an apparent equilibrium constant is only valid for a given pH and cation concentration, and K' must be qualified by information about these conditions. As the standard Gibbs energy change is derived directly from the equilibrium constant, this parameter, when calculated from the apparent equilibrium constant, must be similarly qualified ($\Delta G^{0'}$ $_{(pH=x, [Mg^{2+}]=y)}$).

Calculation of the actual Gibbs energy change requires that the appropriate $\Delta G^{0'}|$ or K' is selected for the prevailing conditions, and that the observed mass–action ratio is calculated using the same conventions:

$$\Delta G' = \Delta G^{0'} + 2 \cdot 3 RT \log_{10} \Gamma' \qquad \text{(Eq. 3.11)}$$

It is frequently, and misleadingly, supposed that the phosphate anhydride bonds of ATP are "high-energy" bonds which are capable of storing energy and driving reactions in otherwise unfavourable directions. However, it should be clear from Table 3.1 that it is the extent to which the observed mass-action ratio is displaced from equilibrium which defines the capacity of the reactants to do work, rather than any attribute of a single component. A hypothetical cell could utilize any reaction to transduce energy from the mitochondrion. For example if the glucose 6-phosphatase reaction were maintained ten orders of magnitude away from equilibrium, then glucose 6-phosphate would be just as capable of doing work in the cell as is ATP. Conversely, the Pacific Ocean could be filled with an equilibrium mixture of ATP, ADP and Pi, but the ATP would have no capacity to do work.

A second error, which still finds its way into print, is the confusion of actual and standard Gibbs energy changes. Standard Gibbs energy changes are merely restatements of the equilibrium constant (Eq. 3.6) and give no information as to the conditions in the cell. The "efficiency" of mitochondrial ATP synthesis is frequently misquoted in this way.

The Gibbs energy for the ATP synthetase reaction is sometimes referred to by the short-hand "phosphorylation potential", "phosphate potential" or ΔG_p. To facilitate a thermodynamic comparison between the phosphorylation potential and redox or proton electrochemical potentials, ΔG_p is frequently expressed in units of millivolts (Section 3.8).

3.3 OXIDATION–REDUCTION (REDOX) POTENTIALS

Both the mitochondrial respiratory chain and the photosynthetic electron transfer chains operate as a sequence of reactions in which electrons are transferred from one component to another. While many of these components simply gain one or more electrons in going from the oxidized to the reduced form, in others the gain of electrons induces an increase in the pK of one or more ionizable groups on the molecule, with the result that reduction is accompanied by the gain of one or more protons.

cytochrome c undergoes a 1 e$^-$ reduction:

$$Fe^{3+} - cyt\ c + 1\ e^- = Fe^{2+} - cyt\ c \qquad \text{(Eq. 3.12)}$$

NAD$^+$ undergoes a 2 e$^-$ reduction and gains one H$^+$:

$$NAD^+ + 2\ e^- + H^+ = NADH \qquad \text{(Eq. 3.13)}$$

while ubiquinone undergoes a 2 e$^-$ reduction followed by the addition of 2 H$^+$:

$$UQ + 2\ e^- + 2\ H^+ = UQH_2 \qquad \text{(Eq. 3.14)}$$

The last is often, but inaccurately, referred to as a 2H-transfer.

All oxidation–reduction reactions can quite properly be described in thermodynamic terms by their Gibbs energy changes. However, as the reactions involve the transfer of electrons, electrochemical parameters are usually employed. Although the thermodynamic principles are the same as for the Gibbs energy change, the origins of oxido-reduction potentials in electrochemistry sometimes obscure this relationship.

The additional facility afforded by an electrochemical treatment of a redox reaction is the ability to dissect the overall electron transfer into two half-reactions, involving respectively the donation and acceptance of electrons.

For example the reaction catalysed by lactate dehydrogenase:

$$Pyruvate + NADH + H^+ \rightleftharpoons Lactate + NAD^+ \qquad \text{(Eq. 3.15)}$$

can be considered as two half-reactions:

$$NADH \rightleftharpoons NAD^+ + H^+ + 2\ e^- \qquad \text{(Eq. 3.16)}$$

$$Pyruvate + 2\ H^+ + 2\ e^- \rightleftharpoons Lactate \qquad \text{(Eq. 3.17)}$$

A reduced oxidized pair such as NADH/NAD$^+$ is termed a redox couple.

As the overall reaction catalysed by lactate dehydrogenase is reversible, each of the half-reactions must in turn be reversible, and so could in theory be described by an equilibrium constant. However, it is not immediately apparent how to treat the electrons, which have no independent existence in solution. A similar problem is encountered in electrochemistry when investigating the equilibrium between a metal (i.e. the reduced form) and a solution of its salt (i.e. the oxidized form). In this case the tendency of the couple to donate electrons is quantitated by forming an electrical cell from two half-cells, each consisting of a metal electrode in equilibrium with a 1 M solution of its salt. An electrical circuit is completed by a bridge which links the solutions without allowing them to mix. The electrical potential difference between the electrodes may then be determined with a high resistance potentiometer. To facilitate comparison of different electrode potentials, they are expressed in relation to a standard electrode assembly, the standard hydrogen electrode:

$$H_2 \rightleftharpoons 2H^+ + 2e$$

Clearly it is not possible to have an electrode of solid hydrogen; instead H_2 gas at 1 atmosphere is bubbled over the surface of a platinum electrode which has been coated with finely divided Pt (platinum black) to increase the surface area. When this electrode is immersed in 1 M H^+ the absolute potential of the electrode is defined as zero (at 25°C). The standard electrode potential of any metal salt couple may now be determined by forming a cell comprising the unknown couple together with the standard hydrogen electrode, or more conveniently with secondary standard electrodes whose electrode potentials are invariant.

A similar approach has been adopted for biochemical redox couples. As with the hydrogen electrode it is not feasible to construct an electrode out of the reduced component of the couple, so a Pt electrode is employed. However, unlike the metal/salt and H_2/H^+ couples both components can generally exist in aqueous solution, and standard conditions are defined in which both the oxidized and reduced components are present at unit activity (or 1 M in concentration terms): note the parallel to the conditions for standard Gibbs energy changes. The experimentally observed potential relative to the hydrogen electrode is termed the standard redox potential.

In only a few cases of bioenergetic relevance do the oxidized and reduced components of the couple equilibrate with the Pt electrode sufficiently rapidly for a stable potential to be registered. In most cases a low concentration of a second redox couple, capable of reacting with both the primary redox couple and the Pt electrode, is added to act as a redox mediator. As will be shown below, the two redox couples achieve equilibrium when they exhibit the same redox potential. As long as the concentration relationships of the primary couple are not disturbed, the electrode will register the potential of the primary couple. The use of one or more mediating couples is of particular importance when the redox potentials of membrane-bound components are being investigated (Section 5.3).

As with the Gibbs energy change, redox potentials are dependent on the relative concentrations of reactants and products. The redox potential (E) at pH $=0$ for the redox couple:

$$ox + ne^- = red$$

is given by the relationship:

ERRATA

p. 51   Equation 3.18 should read   (Eq. 3.18)

$$E = E_0 + 2 \cdot 3 \frac{RT}{nF} \log_{10} \frac{[ox]}{[red]}$$

---

where the standard redox potential E_0 refers to unit concentration of the oxidized and reduced forms. In many cases protons are involved in the redox reaction, in which case the generalized half-reaction becomes:

$$ox + ne^- + mH^+ = red$$

The standard redox potential at a pH other than zero becomes more negative than E_0 at a rate of $2 \cdot 3RT/F$. (m/n) mV per pH unit. This corresponds to -60 mV/pH when $m=n$ and -30 mV/pH when $m=1$ and $n=2$ (Table 3.2). This correction does not apply to the standard hydrogen electrode, which remains a constant reference. The usual bioenergetics convention is to define redox potentials for pH 7. The standard redox potential under these conditions is usually referred to as the mid-point potential $E_{m,7}$. Note that although the E_0 of the hydrogen electrode remains a constant reference, the $E_{m,7}$ for the $H^+/\frac{1}{2} H_2$ couple is $7 \times (-60) = -420$ mV.

Equation 3.18 can now be modified to give the actual redox potential at a pH other than zero:

$$E_{h(pH=x)} = E_{m(pH=x)} + 2 \cdot 3 \frac{RT}{nF} \log_{10} \frac{[ox]}{[red]} \qquad \text{(Eq. 3.19)}$$

The characteristic variation of E_h with the ratio of oxidized to reduced component is shown in Fig. 3.3.

The redox potential enables both redox couples in an oxido-reduction reaction to be discussed independently. However, in order to quantitate the

Tables 3.2   Some mid-point potentials

$ox + ne^- + mH^+ \rightleftharpoons red$

	n	m	$E_{m,7}$ (mV)	Change in E_m when pH increased by 1 unit (mV)
Ferredoxin ox/red	1	0	-430	0
$H^+/\frac{1}{2}H_2$ (H_2 1 atm)	1	1	-420	-60
$NAD^+/NADH$	2	1	-320	-30
$NADP^+/NADPH$	2	1	-320	-30
Menaquinone/menaquinol	2	2	-74	-60
Fumarate/succinate	2	2	$+30$	-60
UQ/UQH_2	2	2	$+40$	-60
Phenazine methosulphate	2	2	$+80$	-60
Cyt c^{3+}/cyt c^{2+}	1	0	$+220$	0
$TMPD^+/TMPD$	1	0	$+260$	0
$Fe(CN)_6^{3-}/Fe(CN)_6^{4-}$	1	0	$+430$	0
Fe^{3+}/Fe^{2+}	1	0	$+780$	0
$O_2/2H_2O^a$	4	4	$+820$	-60

[a] 1 atmosphere O_2 and 55 D (i.e. pure) water

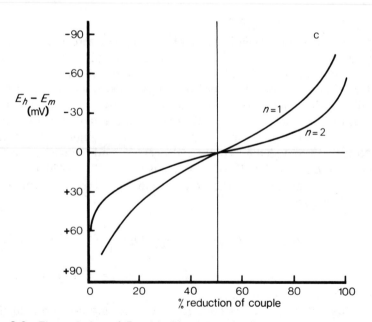

Fig. 3.3   The variation of E_h with the extent of reduction of a redox couple.

thermodynamic disequilibrium between the couples, the redox potential difference ΔE_h is required. For the reaction:

$$A_{red} + B_{ox} = A_{ox} + B_{red}$$

p. 53 Equation 3.20 should read
$$\Delta E_h = E_{h(A)} - E_{h(B)}$$

(Eq. 3.20)

As a complete oxido-reduction reaction can also be described by Gibbs energy changes, Δ_h can also be expressed in kJ mol^{-1}:

$$\Delta G'_0 = -n\mathrm{F}\,\Delta E_h$$

(Eq. 3.21)

The Faraday (F) therefore provides the means of converting from electrical units to kJ mol^{-1} (Table 3.3). From this it is apparent that an oxido-reduction reaction is at equilibrium when $\Delta E_h = 0$.

Table 3.3 Interconversion between redox potential difference and Gibbs energy change for 1-electron and 2-electron transfers

	$\Delta G'$ (kJ mol^{-1})	
ΔE_h (mV)	$n=1$	$n=2$
0	0	0
+100	−9·6	−19·3
+200	−19·3	−38·6
+500	−48·2	−96·5
+1000	−96·5	−193
+1200	−116	−231

Redox potentials were devised for the study of oxido-reduction reactions in a single aqueous phase; for energy-transducing membranes (Section 5.3), however, up to three phases have to be considered: the two aqueous compartments and the membrane itself. This introduces a particular complication if a membrane potential exists across the membrane, since depending on the location of the redox couple the "electrochemical potential of the electron" will include a contribution by the membrane potential (Mitchell 1976a, Walz 1979).

3.4 ION ELECTROCHEMICAL POTENTIAL DIFFERENCES

The third manifestation of the Gibbs energy change in bioenergetics is the potential which is inherent in a gradient of ions across a membrane, in particular a gradient of protons across an energy-transducing membrane. There are two forces acting on an ion gradient across a charged membrane. One is due to the concentration gradient of the ion; the Gibbs energy change for the transfer of one mol of solute from a concentration $[X]'$ to a concentration $[X]''$ in the absence of an electrical potential is given by:

$$\Delta G = 2 \cdot 3RT \log_{10} \frac{[X]''}{[X]'} \qquad \text{(Eq. 3.22)}$$

Note that this equation is closely analogous to that for scalar reactions (Eq. 3.4). In particular the Gibbs energy change is $5 \cdot 7 \text{ kJ mol}^{-1}$ for each tenfold displacement from equilibrium.

The second special case is for the transfer of an ion down an electrical potential difference in the absence of a concentration gradient. In this case the Gibbs energy change when one mol of an ion X^{m+} is transported down an electrical potential of $\Delta\Psi$ mV is given by:

$$\Delta G = -m\text{F}\Delta\Psi \qquad \text{(Eq. 3.23)}$$

In the general case, the ion will be affected by *both* concentrative *and* electrical gradients, and the net Gibbs energy change when one mol of X^{m+} is transported down an electrical potential of $\Delta\Psi$ mV from a concentration of $[X^{m+}]'$ to $[X^{m+}]''$ is given by the general electrochemical equation:

$$\Delta G = -m\text{F}\Delta\Psi + 2 \cdot 3RT \log_{10} \frac{[X^{m+}]''}{[X^{m+}]'} \qquad \text{(Eq. 3.24)}$$

The Gibbs energy change is usually expressed as the ion electrochemical potential gradient $\Delta\tilde{\mu}_{Xm+}$ in units of electrical potential:

$$\Delta\tilde{\mu}_{Xm+} = m\Delta\Psi - \frac{2 \cdot 3RT}{\text{F}} \log_{10} \frac{[X^{m+}]''}{[X^{m+}]'} \qquad \text{(Eq. 3.25)}$$

In the specific case of the proton gradient the logarithmic factor simplifies to the pH difference (ΔpH), so that:

$$\Delta\tilde{\mu}_{H+} = \Delta\Psi - \frac{2 \cdot 3RT}{\text{F}}\Delta\text{pH} \qquad \text{(Eq. 3.26)}$$

Or in mV at 30°C:

$$\Delta\tilde{\mu}_{H+} = \Delta\Psi - 60\Delta\text{pH} \qquad \text{(Eq. 3.27)}$$

While Eq. 3.25 is the most commonly encountered form of the ion electrochemical gradient, it is only valid for ions which cross the membrane by electrical uniport (Section 2.3). When two or more ion movements are directly coupled, it is necessary to consider the electrochemical potential gradient of the complete process. As an example in Fig. 3.4 the electrochemical potential equations for three possible modes of Ca^{2+} transport are considered.

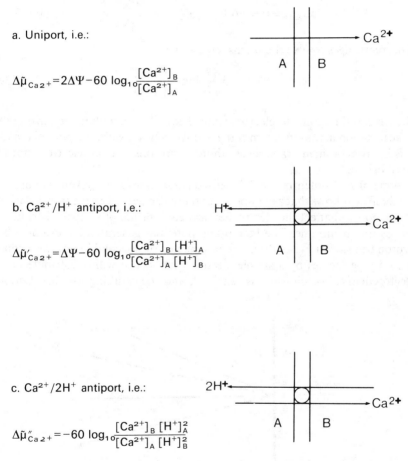

a. Uniport, i.e.:

$$\Delta\tilde{\mu}_{Ca2+}=2\Delta\Psi-60\ \log_{10}\frac{[Ca^{2+}]_B}{[Ca^{2+}]_A}$$

b. Ca^{2+}/H^+ antiport, i.e.:

$$\Delta\tilde{\mu}'_{Ca2+}=\Delta\Psi-60\ \log_{10}\frac{[Ca^{2+}]_B\ [H^+]_A}{[Ca^{2+}]_A\ [H^+]_B}$$

c. $Ca^{2+}/2H^+$ antiport, i.e.:

$$\Delta\tilde{\mu}''_{Ca2+}=-60\ \log_{10}\frac{[Ca^{2+}]_B\ [H^+]_A^2}{[Ca^{2+}]_A\ [H^+]_B^2}$$

Fig. 3.4 The equations of electrochemical potential for Ca^{2+} for different modes of transport.

3.5 EQUILIBRIUM DISTRIBUTIONS OF IONS, WEAK ACIDS AND WEAK BASES

As with all Gibbs energy changes, an ion distribution is at equilibrium across a membrane when ΔG is zero, and hence when $\Delta \tilde{\mu}_{X^{m+}} = 0$. Under these conditions the electrochemical equation becomes:

$$\Delta \tilde{\mu}_{X^{m+}} = 0 = m\Delta\Psi - 2\cdot 3 \frac{RT}{F} \log_{10} \frac{[X^{m+}]''}{[X^{m+}]'} \qquad \text{(Eq. 3.28)}$$

This rearranges to give the Nernst equation:

$$\Delta\Psi = 2\cdot 3 \frac{RT}{mF} \log_{10} \frac{[X^{m+}]''}{[X^{m+}]'} \qquad \text{(Eq. 3.29)}$$

An ion can thus come to electrochemical equilibrium without equalizing its concentration across the membrane. Conversely a membrane potential may exist across a membrane even though protons are at electrochemical equilibrium.

A membrane potential is a delocalized parameter for any given membrane. It therefore follows that a membrane potential generated by the translocation of one ion will affect the electrochemical equilibrium of all ions distributed across the membrane. The membrane potential generated for example by proton translocation can therefore be detected by a second ion. If the second ion only permeates by a simple electrical uniport, it will redistribute until electrochemical equilibrium is regained, and the resulting ion distribution

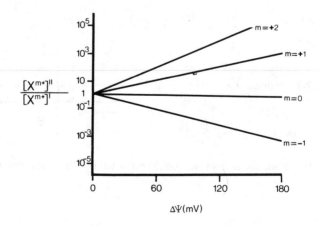

Fig. 3.5   The equilibrium distribution of an ion X^{m+} permeable by uniport across a membrane as a function of membrane potential and the charge carried by the ion.

will enable the membrane potential to be estimated from Eq. 3.29. This is the principle for most determinations of $\Delta\Psi$ in energy-transducing organelles (see Section 4.2). The equilibrium ion distribution varies with $\Delta\Psi$ as shown in Fig. 3.5.

An electroneutrally permeant species will be unaffected by $\Delta\Psi$ and will come to equilibrium when its concentration gradient is zero. Weak acids and bases (i.e. those with a pK' between 3 and 11) can often permeate in the uncharged form across bilayer regions of the membrane (Section 2.4) while the ionized form remains impermeant, even though it may be present in great excess over the neutral species. However, while the neutral species (protonated acid or deprotonated base) equilibrates without regard to $\Delta\Psi$, if a pH gradient exists across the membrane, the Henderson–Hasselbalch equation requires that the concentration of the ionized species must differ for that of the uncharged species to be the same across the membrane (Fig. 3.6). In

a. Weak acids

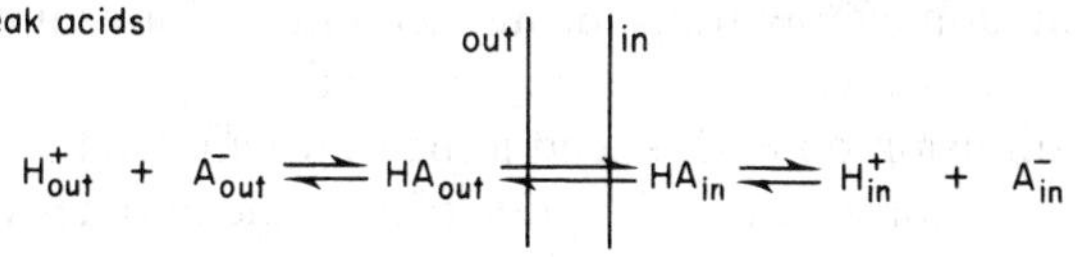

$$H^+_{out} + A^-_{out} \rightleftharpoons HA_{out} \quad | \quad | \quad HA_{in} \rightleftharpoons H^+_{in} + A^-_{in}$$

out | in

If pK of acid is the same in both compartments:

$$K = \frac{[H^+]_{out}\,[A^-]_{out}}{[HA]_{out}} = \frac{[H^+]_{in}\,[A^-]_{in}}{[HA]_{in}}$$

At equilibrium $[HA]_{out} = [HA]_{in}$

$$\therefore [H^+]_{in}/[H^+]_{out} = [A^-]_{out}/[A^-]_{in}$$

b. Weak bases

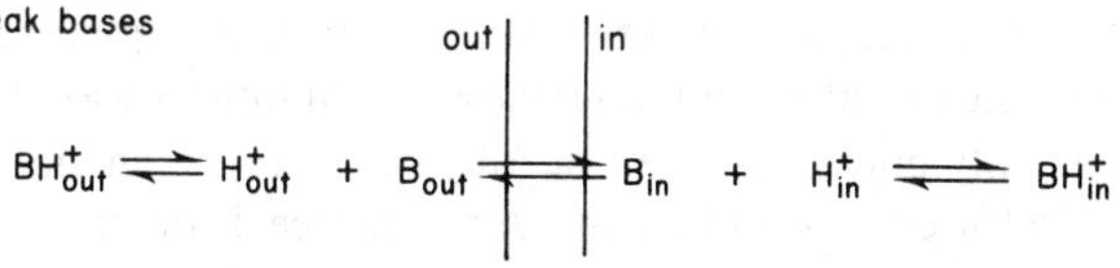

out | in

$$BH^+_{out} \rightleftharpoons H^+_{out} + B_{out} \quad | \quad | \quad B_{in} + H^+_{in} \rightleftharpoons BH^+_{in}$$

$$K = \frac{[H^+]_{out}\,[B]_{out}}{[BH^+]_{out}} = \frac{[H^+]_{in}\,[B]_{in}}{[BH^+]_{in}}$$

At equilibrium $[B]_{out} = [B]_{in}$

$$\therefore [H^+]_{in}/[H^+]_{out} = [BH^+]_{in}/[BH^+]_{out}.$$

Fig. 3.6 The equilibrium distribution of electroneutrally permeant weak acids and bases as a function of ΔpH.

summary, weak acid anions become concentrated in the acidic compartment, while protonated bases concentrate in the alkaline compartment. This principle is widely used to determine ΔpH across energy-transducing membranes (see Section 4.2).

3.6 MEMBRANE POTENTIALS, DIFFUSION POTENTIALS, DONNAN POTENTIALS AND SURFACE POTENTIALS

There are three ways in which a true, bulk-phase membrane potential (i.e. trans-membrane electrical potential difference) may be generated. The first is by the operation of an electrogenic ion pump such as operates in energy-transducing membranes. The second is by the addition to one side of a membrane of a salt, the cation and anion of which have unequal permeabilities. The more permeant species will tend to diffuse through the membrane ahead of the counter-ion and thus create a diffusion potential. Diffusion potentials may be created in energy-transducing membranes for example by the addition of K^+ in the presence of valinomycin (see Fig. 4.7). In energy-transducing organelles, diffusion potentials tend to be transient due to the movement of counter-ions, in contrast to eukaryotic plasma membranes where the generally slow transport processes enable potentials to be maintained for several hours.

The limiting case of a diffusion potential occurs when the counter-ion is completely impermeant. This condition pertains in energy-transducing organelles due to the "fixed" negative charges of the internal proteins and phospholipids. As a result, when the organelles are suspended in a medium of low ionic strength, the more mobile cations attempt to leave the organelle until the induced potential balances the cation concentration gradient. This is a stable Donnan potential.

Donnan and diffusion potentials are both true bulk-phase potentials which may be detected by any technique. However, it must always be borne in mind that $\Delta\Psi$ is only one component of $\Delta\tilde{\mu}_{H+}$. Failure to remember this has been responsible for confusion in some quarters (see Tedeschi 1979). Thus mitochondria suspended in sucrose in the presence of valinomycin (Section 2.5) and FCCP (Section 2.5) maintain a stable Donnan $\Delta\Psi$ of more than 80 mV, although $\Delta\tilde{\mu}_{H+}$ is zero, due to a compensating pH gradient, acid in the matrix (Nicholls 1974). In the presence of substrate and the absence of FCCP, $\Delta\Psi$ undergoes a modest increase, to 150 mV, while $\Delta\tilde{\mu}_{H+}$ increases from zero to 230 mV. The discussion over the extent to which $\Delta\Psi$ is metabolically dependent (Tedeschi 1979) is therefore not relevant to chemiosmosis, whereas the question whether there is a metabolically dependent $\Delta\tilde{\mu}_{H+}$ is crucial.

Surface potentials are quite distinct from the above. Due to the presence of fixed negative charges on the surfaces of energy-transducing membranes, the proton concentration in the immediate vicinity of the membrane is higher than in the bulk phase (Fig. 3.7). However, $\Delta\tilde{\mu}_{H^+}$ is not affected, since the increased proton concentration is balanced by a decreased electrical potential. The proton electrochemical potential difference across the membrane, $\Delta\tilde{\mu}_{H^+}$, is thus unaffected by the presence of surface potentials, although membrane-bound indicators of $\Delta\Psi$, such as the carotenoids of photosynthetic membranes (Section 6.3) will be influenced.

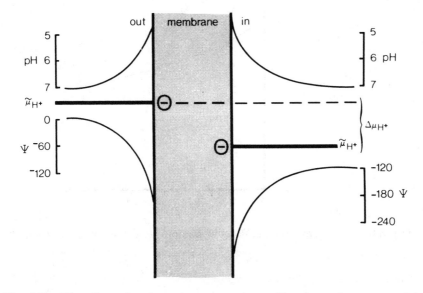

Fig. 3.7 The effect of surface charges on the profile of membrane potential, ΔpH and $\Delta\tilde{\mu}_{H^+}$ in the vicinity of the membrane.

Note that although $\Delta\Psi$ and ΔpH both change close to the membrane, the proton electrochemical potential is unaffected by surface charge. This means that a proton close to, or even bound to, the surface of the membrane is in electrochemical equilibrium with the bulk phases. (After Junge, 1977.)

3.7 PHOTONS

In photosynthetic systems, the primary source of Gibbs energy is the quantum of electromagnetic energy, or photon, which is absorbed by the photosynthetic pigments. The energy in a single photon is given by $h\nu$, where h is Planck's constant ($6{\cdot}62 \times 10^{-34}$ J s) and ν is the frequency of the radiation

(s^{-1}). One photon interacts with one molecule, and therefore N photons, where N is Avogadro's constant, will interact with one mol. The energy in one mol (or einstein) of photons is therefore:

$$E = Nhv = Nhc/\lambda = 120\ 000/\lambda \text{ kJ einstein}^{-1} \qquad \text{(Eq. 3.30)}$$

where c is the velocity of light, and λ is the wavelength in nm. As may be seen in Fig. 3.8, even the absorption of an einstein of red light (600 nm) makes available 200 kJ mol^{-1} which compares favourably with the Gibbs energy changes encountered in bioenergetics.

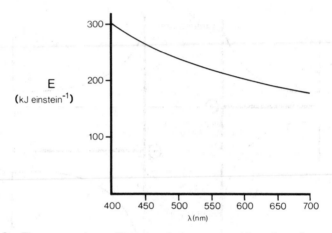

Fig. 3.8   The energy in an Einstein of photons as a function of wavelength.

3.8 BIOENERGETIC INTERCONVERSIONS

The critical stages of chemiosmotic energy transduction involve the inter-conversions of Gibbs energy changes between the different forms discussed in the previous sections. Some of the more important interconversions are summarized in Table 3.4. Equilibrium thermodynamics is competent to answer the following questions. (i) Is a given interconversion at equilibrium $(\Delta G = 0)$? (ii) In which direction will an interconversion run? (iii) What is the stoicheiometry of a given interconversion (that so that ΔG may be calculated to be zero when equilibrium is known to be attained)? In addition, as will be seen below (Section 3.9) the parameters of equilibrium thermodynamics have a limited applicability within the formalism of irreversible thermodynamics, since for near-equilibrium processes the rate of energy

Table 3.4 Equilibrium conditions for some chemiosmotic energy transductions

---

(*a*) REDOX-DRIVEN PROTON PUMP
$$2\Delta E_h = n\Delta\tilde{\mu}_{H^+}$$

where n is the $H^+/2e^-$ stoicheiometry for the redox span under consideration. Note that this equation is only valid if the two redox couples defining the redox span are located in the same compartment (Walz 1979)

(*b*) ATP SYNTHETASE
$$\Delta G_{p(in)} = n'\Delta\tilde{\mu}_{H^+}$$
$$\Delta G_{p(out)} = (n'+1)\ \Delta\tilde{\mu}_{H^+}$$

where n' is the H^+/ATP stoicheiometry, $\Delta G_{p(in)}$ refers to the phosphorylation potential in the mitochondrial matrix, and $\Delta G_{p(out)}$ refers to the phosphorylation potential in the extra-mitochondrial compartment, the difference being due to the transport of adenine nucleotides and Pi across the inner membrane (Section 7.6)

---

flux is a function of the thermodynamic disequilibrium calculated from equilibrium values.

To compute the overall Gibbs energy change for a given interconversion, it is necessary to express each component in the same units. While the Gibbs energy change (in kJ mol^{-1}) is the more fundamental unit, redox potentials and electrochemical potential gradients are more familiar in units of electrical potential, and it is therefore common to convert Gibbs energy changes of non-redox reactions into electrical units by dividing by the Faraday constant:

$$\Delta G/F = \Delta G^{0'}/F + 2.3RT/F.\log_{10}\Gamma' \qquad \text{(Eq. 3.31)}$$

Thus if the observed mass-action ratio of a reaction is one order of magnitude away from equilibrium (Eq.3.4) then the Gibbs energy change of 5·7 kJ mol^{-1} corresponds to 59 mV. In Table 3.4 the equilibrium condition is defined for a number of interconversions.

While bioenergetic systems are open *in vivo*, with isolated organelles it is frequently possible to allow a given interconversion to achieve a true equilibrium by the simple expedient of inhibiting subsequent steps. For example, isolated mitochondria can achieve an equilibrium between $\Delta\tilde{\mu}_{H^+}$ and the $ATP/ADP+Pi$ reaction by avoiding any extra-mitochondrial ATPase activity. Under such conditions a thermodynamically defined stoicheiometry can be calculated (Section 4.4).

3.9 THE APPLICATION OF IRREVERSIBLE THERMODYNAMICS

Reviews Rottenberg 1979b, Westerhoff & van Dam 1979

Classic equilibrium thermodynamics is valid for predicting the conditions under which a given energy transducing step would achieve equilibrium and by extension for predicting the direction, but not the rate, in which the process might be expected to proceed when that equilibrium is disturbed. Non-equilibrium (or irreversible) thermodynamics is much less restricted; it can relate the net rate of energy transduction to the extent of thermodynamic disequilibrium and can in addition be used to build up a set of equations from which the flows, potentials and mutual interactions of electron transfer, proton translocation and ATP synthesis can be derived. The flow of electrons in an electrical circuit can be described by similar approach, which is why the electrical analogy of the proton circuit (Section 4.1) is valid, but restricted to the one dimension of the proton flow.

It is beyond the scope of this book to describe the formalism of irreversible thermodynamics. Nevertheless it is important to appreciate the areas in which the treatment should be applied.

Under near-equilibrium conditions, the net flux J through a bioenergetic process is a function of A, the affinity or Gibbs energy difference. Under near-equilibrium conditions, flux and affinity are linearly related by a co-efficient L, thus:

$$J = L.A$$

In the case of the transport of an ion across a membrane by a simple uniport mechanism, the affinity is equivalent to the electrochemical potential difference (expressed in the same units as the affinity):

$$J = L.\Delta\tilde{\mu}$$

For example, in the case of a leak catalysed by a proton translocator:

$$J_{H+} = L_{H+}.\Delta\tilde{\mu}_{H+}$$

i.e. the leakage flux of protons should be linearly related to $\Delta\tilde{\mu}_{H+}$. The equations remain valid for large deviations from equilibrium in the case of diffusion processes, but not in the case of chemical reactions.

For a tightly coupled symport of two ions across a membrane:

$$J_B^A = L(\Delta\tilde{\mu}_A + \Delta\tilde{\mu}_B)$$

For example, for the proton:lactose symport catalysed by the *lac* permease of *E. coli* (Section 8.5):

$$J^{H^+}{}_{lac} = L(\Delta\tilde{\mu}_{lac} + \Delta\tilde{\mu}_{H+})$$

This implies that lactose can be transported up its electrochemical potential gradient, as long as $\Delta\tilde{\mu}_{H+}$ is sufficient.

For an ion pump with a stoicheiometry of n, for example the ATP synthetase, the ion and chemical fluxes are related thus:

$$J_{ATP} = L_{ATP} (A_{ATP} + n.\Delta\tilde{\mu}_{H+})$$

$$J_{H+} = n.J_{ATP}$$

In other words, the net rate of energy transduction by an ion pump is proportional to the overall thermodynamic disequilibrium.

For the mitochondrion, with proton fluxes attributable to the respiratory chain, the ATP synthetase and the leak proton conductance of the membrane can be described by similar equations, as can the flux of electrons down the respiratory chain and the flux of ATP synthesis. Each flux is equal to the product of a proportionality constant and the Gibbs energy difference of the step.

The most important application of these equations is in the attempt to derive the true mechanistic stoicheiometries of proton translocation and ATP synthesis from experiments where, almost inevitably, extraneous leaks are occurring (see Van Dam et al. 1980).

The proton circuit of brown adipose tissue (BAT) mitochondria has a leak which can be plugged by nucleotides such as GDP. This is part of the thermogenic mechanism enabling fatty acids to be oxidized to acetate for heat. The Cl^- permeability is puzzling (Section 4.5)

4 The Chemiosmotic Proton Circuit

4.1 INTRODUCTION

The circuit of protons linking the primary generators of $\Delta\tilde{\mu}_{H^+}$ with the ATP synthetase was introduced in Chapter 1. The purpose of this chapter is to discuss the function of the proton circuit in the wide range of chemiosmotic energy transductions which can be created *in vitro*. The close analogy between the proton circuit and the equivalent electrical circuit (see Fig. 1.3) will be emphasized, not only as a simple model but also because the same laws govern the flow of energy around both circuits (Section 3.9).

In an electrical circuit the two fundamental parameters are potential difference (in volts) and current (in amps). From measurements of these functions other factors may be derived, such as the rate of energy transmission (in watts) or the resistance of components in the circuit (in ohms). In Fig. 4.1 a simple electrical circuit is shown, together with the analogous proton circuit across the mitochondrial inner membrane (the circuit operating across a photosynthetic membrane would be closely analogous). In an open circuit (Fig. 4.1a), electrical potential is maximal, but no current flows as the redox potential difference generated by the battery is precisely balanced by the back-pressure of the electrical potential. The tight coupling of the redox reactions within the battery to electron flow prevents any net chemical reaction. In the case of the mitochondrion, the proton circuit is open-circuited when there are no means for the protons extruded by the respiratory chain to re-enter the matrix. As with the electrical circuit the potential across the membrane is maximal under these conditions, and the redox potential difference across the proton-translocating regions of the respiratory chain (Section 5.3) is in equilibrium with the proton electrochemical potential (allowing for the H^+/e^- stoicheiometry of the interconversion (Section 3.8)). As the redox reactions are tightly coupled to proton extrusion there is no respiration in this condition.

In Fig. 4.1b the electrical and proton circuits are shown operating normally and performing useful work. The potential is slightly less than under open-circuit conditions, as the net driving force enabling the battery and respiratory

MITOCHONDRION   ELECTRICAL ANALOGUE

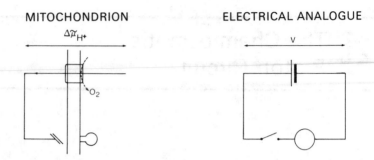

a. Open circuit. Current zero (no respiration). Potential ($\Delta\tilde{\mu}_{H^+}$) is maximal.

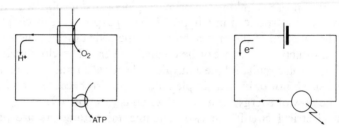

b. Circuits completed, current flows (respiration occurs). Useful work is done (ATP is synthesized). Potential ($\Delta\mu_{H^+}$) is less than maximal.

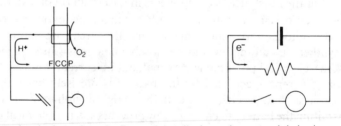

c. Short-circuit introduced. Energy is dissipated, potential is low, current (respiration) is high.

Fig. 4.1   The mitochondrial proton circuit is analogous to an electrical circuit.

chain to operate is the slight disequilibrium between the redox potential difference available and the potential in the circuit. The "internal resistance" of the battery may be calculated from the drop in potential required to sustain a given current. Analogously, the "internal resistance" of the respiratory chain may be estimated, and is found to be very low (Section 4.5).

An electrical circuit may be shorted by introducing an additional low resistance pathway in parallel with the existing circuit (Fig. 4.1c). Current can now flow from the battery without having to do useful work, the energy being dissipated as heat. This uncoupling of the circuit can be accomplished in the proton circuit by the addition of proton translocators (Section 2.4), enabling respiration to occur without stoicheiometric ATP synthesis.

It is an over-simplification to consider the respiratory chain as a single generator of "proticity", as the chain consists of three proton pumps operating in parallel from the standpoint of the proton circuit and in series with respect to the electron flow (Fig. 4.2). Detailed discussion of the respiratory chain will be reserved for Chapter 5. Here they will be treated merely as "black boxes", although it should be noted that it is possible to introduce or remove electrons at the interface of each proton pump, thus enabling each pump to be studied in isolation.

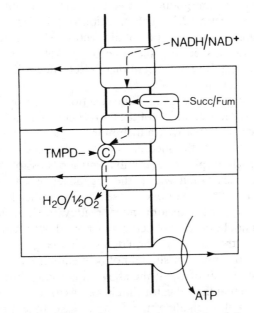

Fig. 4.2 The mitochondrial respiratory chain consists of three proton pumps which act in parallel with respect to the proton circuit and in series with respect to the electron flow.

4.2 THE MEASUREMENT OF PROTON ELECTROCHEMICAL POTENTIAL

Reviews Rottenberg 1975, 1979a, Fillingame 1980

Quantification of the proton electrochemical potential has been important for the chemiosmotic theory, since a single demonstration of net ATP synthesis in the absence of a sufficient $\Delta\tilde{\mu}_{H^+}$ would be sufficient to demolish the entire edifice. The proton electrochemical potential is also the most direct and quantitative indicator of the state of "energization" existing in the energy-transducing organelle.

All techniques for the determination of $\Delta\tilde{\mu}_{H^+}$ involve the separate estimation of $\Delta\Psi$ and ΔpH (Section 3.5). $\Delta\Psi$ is calculated either by direct determination of the concentration gradient at equilibrium of an ion permeable by electrical uniport (Section 2.3) by application of the Nernst equation (Eq. 3.30), or by using the diffusion potential (Section 3.6) generated by an ion gradient to calibrate a spectroscopic indicator of $\Delta\Psi$. ΔpH is generally calculated from the equilibrium distribution of electroneutrally permeant weak acids and bases (Section 3.5).

Considerable care must be taken in the selection of appropriate indicators. To measure $\Delta\Psi$ the ion should be of the correct charge to be accumulated (a cation if the interior is negative (mitochondria) or an anion if the interior is positive with respect to the medium (sonic sub-mitochondrial particles, chromatophores, chloroplasts). Secondly, it must be possible to calculate the free indicator concentration within the organelle, by choosing either an indicator which is not bound or one whose activity coefficient may be readily calculated. Thirdly, the indicator must readily achieve electrochemical equilibrium and not be capable of being transported by more than one mechanism. Fourthly, the indicator should disturb the gradients as little as possible. Fifthly, the indicator should not be metabolized.

For the measurement of ΔpH, the above conditions should be satisfied, with the exception that the indicator should be a weak acid to be accumulated within organelles with an alkaline interior, and a weak base to be accumulated in an acidic compartment (Section 3.5).

Once an equilibrium distribution of an indicator has been achieved it must be measured. This can be accomplished by rapid separation of the organelle from the incubation medium, by continuously monitoring the fall in indicator concentration in the incubation as the ion is accumulated, or by making use of the altered spectral properties of the indicator when accumulated within an organelle. Examples of these techniques will now be considered.

4.2.1 The determination of $\Delta\mu_{H^+}$ by ion-specific electrodes

Reviews Rottenberg 1975, 1979a, Skulachev 1979

The first determination of $\Delta\tilde{\mu}_{H+}$ in mitochondria (Mitchell & Moyle 1969a) employed pH- and K^+-specific electrodes in an anaerobic mitochondrial incubation (Fig. 4.3). Valinomycin was present to create a high electrical uniport for K^+, and $\Delta\Psi$ was calculated from the K^+ uptake occurring on the addition of an extended pulse of O_2. ΔpH was estimated from the parallel H^+-extrusion. A value of 228 mV was obtained for $\Delta\tilde{\mu}_{H+}$ respiring under "open circuit" conditions in the absence of ATP synthesis (State 4 in Table 4.1).

Table 4.1 Respiratory states[a]

				Respiration
State 1	mitochondria	no substrate	no ADP	low
State 2	mitochondria	no substrate	ADP	low
State 3	mitochondria	substrate	ADP	high
State 4	mitochondria	substrate	ADP exhausted	low
State 5	mitochondria	substrate	oxygen exhausted	

State 3 can be sub-divided
   State 3_{unc}: high respiration achieved by addition of proton translocator
   State 3_{ADP}: high respiration achieved by ADP addition.
   State $3\frac{1}{2}$: intermediate state with ADP supply rate-limiting (physiological)

State 6 refers to the inhibited respiration sometimes observed when a high ΔpH is built up as a result of Ca^{2+} accumulation in the absence of permeant anion

[a] The original classification of respiratory states (Chance & Williams 1956) followed the order of additions in an oxygen electrode experiment such as that shown in Fig. 4.17

A closely related method was developed in Skulachev's laboratory to measure $\Delta\Psi$ in a variety of energy-transducing organelles. In place of K^+ *plus* valinomycin, a series of synthetic cations and anions were synthesized whose charge was sufficiently delocalized and screened with hydrophobic groups to enable the charged species to permeate bilayer regions (Section 2.5). In the original method, the fall in concentration of the ion in the incubation medium due to uptake into the organelle was detected by a planar black

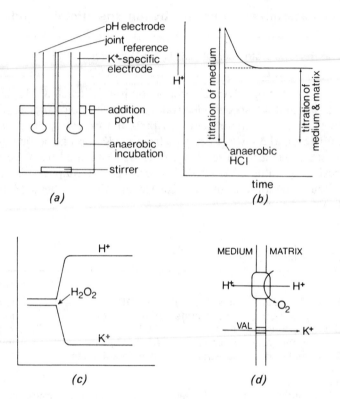

Fig. 4.3 The estimation of $\Delta\Psi$ and ΔpH by ion-specific electrodes.

(a) Apparatus; (b) determination of matrix buffering capacity; (c) experiment;
(d) ion movements.

Mitochondria were incubated in an anaerobic medium containing sucrose, substrate, valinomycin and a low concentration of K^+. Sufficient H_2O_2 was then added to allow the respiratory chain to function for about 3 min (the incubation contained catalase). Protons were expelled (measured by the pH-electrode), and K^+ taken up (measured by the K^+-specific electrode). K^+ distributed according to the Nernst equation in the presence of valinomycin, and the K^+ gradient was calculated from the fall in external K^+, knowing the volume of the matrix into which the K^+ was being accumulated. To calculate ΔpH from the fall in external pH, the rise in matrix pH due to the loss of protons from the matrix is needed. This in turn requires the matrix buffering capacity. This is obtained in an independent experiment in which an anaerobic pulse of HCl is added. The initial acidification is a measure of the buffering capacity of the medium. There is then a partial decay as protons enter the matrix, the final state reflecting the buffering capacity of medium+matrix.

lipid membrane (Section 1.3) separated from a second compartment also containing the "Skulachev" cation. The difference in ion concentration across the synthetic membrane created a membrane potential which could be detected by electrodes and thus acted as an ion-specific electrode. This technique, by avoiding the use of K^+ and valinomycin minimized the "clamping" of $\Delta\Psi$ inherent in the use of an ion which was already present at 100 mM in the matrix (see below), and caused the final demise of the chemical intermediate cation-pump (Section 1.4).

A number of these ions are now available as radio-isotopes, and can therefore be used in isotopic determinations of $\Delta\Psi$.

4.2.2 The estimation of $\Delta\Psi$ and ΔpH by isotope distribution

---

Reviews Nicholls 1974, Rottenberg 1975, 1979, Ramos *et al.* 1979

It is possible to determine $\Delta\Psi$ and ΔpH from equilibrium radio-isotope distributions without the necessity of separating the organelles from the incubation by the techniques of flow dialysis (Fig. 4.4). The concentration of the indicator in the incubation is monitored continuously by its rate of diffusion across a semipermeable membrane into a steady flow of medium. However, while avoiding the need to separate the organelles, this method shares a limitation of sensitivity with the electrode techniques discussed above. In most incubations, the volume of the internal compartment is about three orders of magnitude smaller than that of the incubation. It follows that a method which directly detects an increase in ion concentration in the internal compartment, rather than a fall in concentration in the incubation, will be much more sensitive, even though separation is required.

A successful separation technique must be rapid in order to avoid artifactual redistribution of the indicators, and must provide a means of estimating any residual contamination with incubation medium. Two techniques have been generally applied: silicone oil centrifugation (Fig. 4.5) and millipore filtration. Either technique has its advantages. Oil centrifugation is precise (due to a low contamination with incubation) but time-consuming. Filtration is less precise (due to limitations of loading on the filter and a high external volume) but extremely rapid and simple. In either case a second isotope must be included to act as an external marker for the quantification of the external contamination. Table 4.2 lists some isotopic combinations which have been employed in the determination of $\Delta\Psi$ and ΔpH.

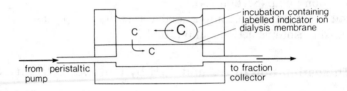

incubation containing
labelled indicator ion
dialysis membrane

C

C

C

from peristaltic
pump

to fraction
collector

b. Experiment

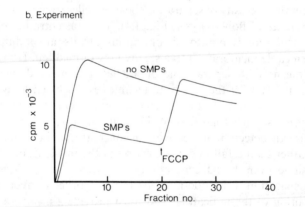

cpm × 10⁻³

10

5

no SMPs

SMPs

FCCP

10   20   30   40

Fraction no.

Fig. 4.4 Determination of $\Delta\Psi$ in sub-mitochondrial particles by flow-dialysis. The upper chamber of the flow-dialysis cell contained: 20µM S¹⁴CN (as electrically permanent anion); NAD⁺, ethanol and alcohol dehydrogenase (to act as an NADH-regenerating system to provide substrate); and acetate to limit the build-up of ΔpH. Sub-mitochondrial particles (SMPs) were added where shown. Uptake of S¹⁴CN into the SMPs lowered the isotope concentration in the incubation, with the result that the rate of diffusion across the dialysis membrane was reduced in strict proportion. Addition of FCCP collapsed $\Delta\Psi$, inducing an efflux of isotope from the SMP, with the result that diffusion across the dialysis membrane increased (data from Sorgato *et al.* 1978).

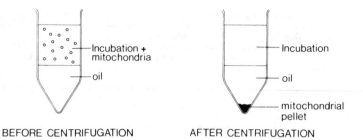

BEFORE CENTRIFUGATION   AFTER CENTRIFUGATION

Fig. 4.5 Silicone-oil centrifugation of mitochondria for the determination of $\Delta\Psi$. Mitochondria are incubated under the desired conditions in a K^+-free medium containing valinomycin, with $^{86}Rb^+$, $[^{14}C]$-sucrose and ^3H_2O. An aliquot is added to a small centrifuge tube containing sislicone oil and is centrifuged at about 10 000g for 1 min. The mitochondria form a pellet under the oil and can be solubilized for liquid scintillation counting. An aliquot of the supernatant is also counted. The $[^{14}C]$-sucrose in the incubation allows the sucrose-permeable space (V_s) in the pellet to be calculated. This gives the extra-matrix contamination of the pellet with incubation medium. The difference between the ^3H_2O-permeable space (V_H) and V_s gives the sucrose-*im*permeable space, which is taken to represent the matrix volume (as water but not sucrose can permeate the inner membrane). If the apparent R_b^+-space in the pellet is calculated (V_{Rb}), then $\Delta\Psi$ can be calculated from the Nernst relationship:

$$\Delta\Psi = 59 \log_{10}\frac{(V_{Rb}-V_s)}{(V_H-V_s)}$$

4.2.3 Optical indicators of $\Delta\Psi$ and ΔpH

Reviews Waggoner 1976, Bashford & Smith 1979

A $\Delta\Psi$ of some 200 mV corresponds to an electrical field across the energy-transducing membrane in excess of 300 000 V cm^{-1}. It is not surprising therefore that certain natural membrane constituents respond to the electrical field by altering their spectral properties. Such electrochromism is due to the effect of the imposed field on the energy levels of the electrons in the molecule. The most widely studied of these intrinsic probes of $\Delta\Psi$ are the pigments, in particular the carotenoids, of photosynthetic energy-transducing membranes (Section 6.3). These respond with extreme rapidity (ns or less) and enable the primary electrogenic events of the light reaction to be followed (Section 6.2). One limitation, however, is that being bound in the membrane they only detect the field in their immediate environment, which need not correspond to the bulk-phase membrane potential difference measured by

Table 4.2   Some values for the proton electrochemical potential across energy-transducing membranes

Material	Conditions	Method	$\Delta\Psi$	$-60\Delta pH$	$\Delta\bar{\mu}_{H+}$	Reference
Liver mitochondria	State 4	Ion-specific electrodes	168	48	216	Mitchell & Moyle 1969
Brown fat mitochondria	Proton channel open	Isotope distribution of $^{86}RB_3^+$, 14C-methylamine and H-acetate (Filtration)	79	-25	54	Nicholls 1979
	Proton channel inhibited		134	$+95$	229	
Heart sub-mitochondrial particles	NADH substrate	Isotope distribution of $S^{14}CN$ and ^{14}C-methyl-amine. (Flow dialysis)	145	0	145	Sorgato *et al.* 1978
E. coli cells	Respiring	Tetraphenylphosphonium uptake ($\Delta\Psi$), and DMO (ΔpH) (Flow dialysis)	100	$+105$	205	Zilberstein *et al.* 1979
Chloroplasts	light	ΔpH by 9-aminoacridine fluorescence quenching $\Delta\Psi$ zero	0	180	180	Rottenberg *et al.* 1972
Chromatophores	dark	Isotope distribution of ^{14}CNS and ^{14}C-methyl-amine	12	65	78	Schuldiner *et al.* 1974
	light	(Centrifugation after protamine sulphate aggregation)	89	106	195	

distribution techniques particularly as surface potential effects (Section 3.6) may be significant.

Extrinsic probes of $\Delta\Psi$ and ΔpH have to be added to the incubation. Some examples of molecules which have proved useful are shown in Fig. 4.6. No extrinsic probe has yet been described which acts like the intrinsic caroten-oids, i.e. remaining fixed in the membrane and altering their spectrum in re-sponse to the applied field. Indeed, while extrinsic probes are valuable tools for following $\Delta\Psi$, and ΔpH, their mechanisms are still a matter of some debate. A common feature is that their absorption or fluorescent emission spectra are modified when the probe is bound to, or accumulated within, an organelle. This enables the extent of association to be followed in incuba-tions without the necessity to separate the organelles. Since they are large planar molecules, the probes often possess the ability to form stacks of molecules when in locally high concentration. This stacking reduces their ability to absorb light and underlies at least some of the observed spectral changes. The complex nature of the probe response means that many factors other than $\Delta\Psi$ or ΔpH can interfere, and great care must therefore be taken

A. Cationic
1, Phenosafranine

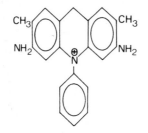

2, Cyanine dyes, e.g.

B. Anionic
1, Oxonols, e.g.

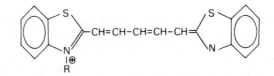

Fig. 4.6 Some optical indicators of membrane potential.

to control the conditions under which they are used. The spectral response must always be calibrated, for example by reference to a diffusion potential (Section 3.6) generated by a known K^+ gradient across the membrane in the presence of valinomycin (Fig. 4.7).

4.2.4 The relative contributions of $\Delta\Psi$ and ΔpH

Review Gromet–Elhanan 1977

The events which regulate the partition of $\Delta\tilde{\mu}_{H+}$ between $\Delta\Psi$ and ΔpH are summarized in Fig. 4.8. Starting from a "de-energized" organelle such as a mitochondrion, the operation of a H^+-pump in isolation leads to the establishment of a $\Delta\tilde{\mu}_{H+}$ in which the dominating component is $\Delta\Psi$. In the case of

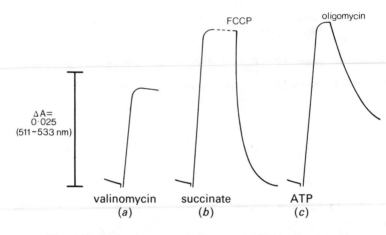

Fig. 4.7   Safranine as an indicator of $\Delta\Psi$ in mitochondria.
Liver mitochondria were incubated in the presence of rotenone to inhibit respiration (Section 5.6). In trace (*a*) a diffusion potential of 124 mV was induced by the addition of valinomycin. The concentration of K^+ in the medium was 0.96 mM, and that in the matrix was assumed to be 120 mM. In trace (*b*), $\Delta\Psi$ was induced by addition of the respiratory substrate succinate and abolished by addition of a proton translocator. In trace (*c*) $\Delta\Psi$ was induced by ATP hydrolysis and abolished by addition of the ATP synthetase inhibitor oligomycin (Section 7.2). 10μM safranine was present in the incubations, which were performed in a dual-wavelength spectrophotometer (Section 5.2). (Data from Åkerman & Wikström 1976.)

the mitochondrion, the electrical capacity of the membrane is such that the net transfer of 1 nmol of H^+ mg protein^{-1} across the membrane establishes a $\Delta\Psi$ of about 200 mV. The pH buffering capacity of the matrix is about 20 nmol of H^+ mg protein^{-1} per pH unit, and the loss of 1 nmol of H^+ will only increase the matrix pH by 0·05 units (i.e. $-60 \Delta pH = 3$ mV). $\Delta\tilde{\mu}_{H^+}$ will thus be about 99% in the form of a membrane potential (Fig. 4.8b).

Following the establishment of a $\Delta\tilde{\mu}_{H^+}$, the second event which might occur is the redistribution of electrically permeant ions (Fig. 4.8c). Uptake for example of K^+ in the presence of valinomycin, or of Ca^{2+} (Section 8.4) will tend to dissipate $\Delta\Psi$ and hence lower $\Delta\tilde{\mu}_{H^+}$. The respiratory chain responds to the lowered $\Delta\tilde{\mu}_{H^+}$ by a further net extrusion of protons, thus increasing ΔpH. Because of the pH buffering capacity of the matrix, the uptake of 20 nmol of K^+ (or 10 nmol of Ca^{2+}), balanced by the extrusion of 20 nmol of H^+ will lead to the establishment of a ΔpH of about 1 unit ($-60 \Delta pH = 60$ mV). As the respiratory chain can only achieve the same total $\Delta\tilde{\mu}_{H^+}$ as before, this means that the final $\Delta\Psi$ must be nearly 60 mV lower than before uptake of the cation. Thus cation uptake leads to a redistribution from $\Delta\Psi$

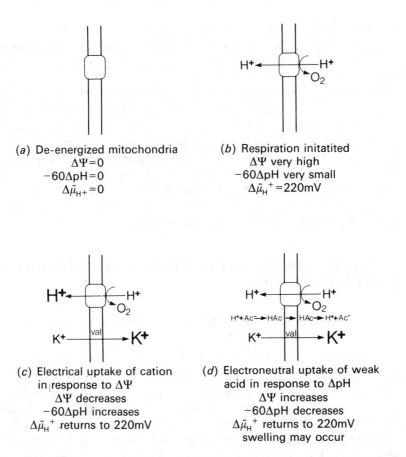

(a) De-energized mitochondria
$\Delta\Psi = 0$
$-60\Delta pH = 0$
$\Delta\tilde{\mu}_{H+} = 0$

(b) Respiration initatited
$\Delta\Psi$ very high
$-60\Delta pH$ very small
$\Delta\tilde{\mu}_H^+ = 220mV$

(c) Electrical uptake of cation
in response to $\Delta\Psi$
$\Delta\Psi$ decreases
$-60\Delta pH$ increases
$\Delta\tilde{\mu}_H^+$ returns to 220mV

(d) Electroneutral uptake of weak
acid in response to ΔpH
$\Delta\Psi$ increases
$-60\Delta pH$ decreases
$\Delta\tilde{\mu}_H^+$ returns to 220mV
swelling may occur

Fig. 4.8 Factors controlling the partition of $\Delta\tilde{\mu}_H^+$ between $\Delta\Psi$ and ΔpH.

to ΔpH. The lowered $\Delta\Psi$ means that cation uptake under these conditions becomes self-limiting, as the driving force steadily decreases until electrochemical equilibrium (Eq. 3.28) is attained. For example the uptake of Ca^{2+} by mitochondria in exchange for extruded protons is limited to about 20 nmol mg protein^{-1} (Section 8.4), by which time $\Delta\Psi$ has decreased (and $-60 \Delta pH$ increased) by about 120 mV.

The third event which can influence the relative contributions of $\Delta\Psi$ and ΔpH is the redistribution of electroneutrally permeant weak acids and bases (Fig. 4.8d). For example, uptake of a weak acid in response to the ΔpH created by prior cation accumulation dissipates ΔpH and allows the respiratory chain to restore $\Delta\Psi$. However, the situation has now been reached where both cation and anion have been accumulated, and if amounts are large, osmotic swelling of the matrix (Section 2.7) results. This does not occur when the cation and anion are respectively Ca^{2+} and Pi, as formation of a non-osmotically active calcium phosphate complex prevents an increase in internal osmotic pressure (Section 8.4).

It is clear from the above discussion that $\Delta\Psi$ and ΔpH indicators themselves, being ions weak acids or weak bases, can disturb the very gradients to be measured unless care is taken. This is particularly true in the presence of valinomycin, as the ionophore brings into play the high endogenous K^+ of the matrix, with the result that $\Delta\Psi$ can become clamped at the value given by the initial K^+ gradient. This risk is less apparent with "Skulachev" cations such as $TPMP^+$ which can be employed at very low concentrations.

Some representative values for $\Delta\Psi$ and ΔpH for a variety of energy-transducing membranes are summarized in Table 4.2.

4.3 THE STOICHEIOMETRY OF PROTON EXTRUSION BY THE RESPIRATORY CHAIN

Reviews Papa 1976, Mitchell *et al.* 1979, Reynafarje *et al.* 1979, Wikström & Krab 1980

The proton current generated by the respiratory chain cannot be determined directly under steady-state conditions as there is no means of detecting the flux of protons when the rate of H^+-efflux exactly balances that of re-entry. It is, however, possible to measure the initial ejection of protons which accompanies the onset of respiration before H^+-re-entry has become established. By making a small precise addition of O_2 to an anaerobic mitochondrial suspension in the presence of substrate and monitoring the extent of H^+-extrusion with a rapidly responding pH-electrode, it is therefore possible to obtain a value for the H^+/O stoicheiometry for the segment of the respiratory chain between the substrate and O_2. The practical details of such an experiment are shown in Fig. 4.9. A number of precautions are necessary in order to obtain accurate results. First, an electrical cation permeability has to exist so that H^+-extrusion can be charge-compensated, preventing the build-up of $\Delta\Psi$ which would otherwise oppose further H^+-extrusion (Section 4.2). Secondly, the pulse of O_2 must be small enough to prevent ΔpH from attaining a saturating value. Thirdly, however rapid the pulse of respiration, the low proton permeability will result in some protons having leaked back across the membrane (and thus being undetected) before the burst of respiration is completed. These protons must therefore be allowed for: the problem is enhanced if Pi (which is nearly always present in mitochondrial prepara-

Experiment

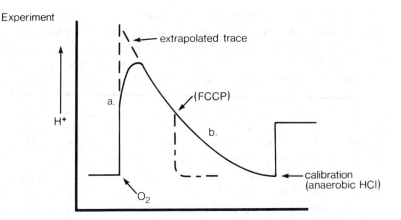

Ion movements

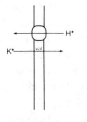

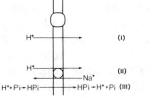

a. aerobic phase

b. anaerobic phase, factors contributing
to the decay of ΔpH.

Fig. 4.9 Determination of mitochondrial H^+/O ratios by the oxygen pulse technique.

The apparatus is the same as that for the determination of $\Delta\bar{\mu}_{H+}$ (Fig. 4.3) except that the K^+-electrode is absent. A concentrated mitochondrial suspension is incubated anaerobically in a lightly buffered medium containing substrate, valinomycin and a *high* concentration of KCl. The pH electrode must have a rapid response. To start the transient a *small* aliquot of air-saturated medium (containing about 5 nmol of O mg protein^{-1}) is rapidly injected. There is a rapid acidification of the medium as the respiratory chain functions for 2–3 s while using up the added O_2. Valinomycin and K^+ are necessary to prevent the build-up of a $\Delta\Psi$ which would inhibit further proton extrusion. When O_2 is exhausted the pH decays as protons leak back into the matrix. This decay can be due to: (i) proton permeability of the membrane, thus FCCP addition during the decay phase accelerates decay; (ii) the endogenous Na^+/H^+ antiport; (iii) electroneutral Pi entry (see Section 7.5). The trace must then be corrected by extrapolation to allow for proton re-entry occurring before the O_2 is exhausted.

tions) is allowed to re-enter the mitochondrion during the O_2-pulse. H/Pi symport (Section 7.6) is extremely active in most mitochondria, and Pi uptake due to the induced ΔpH results in a movement of protons into the matrix and hence an underestimate of H^+/O stoicheiometry. Inhibition of the symport by N-ethylmaleimide significantly increases the observed H^+/O ratio (Brand *et al.* 1976, but see Moyle & Mitchell 1978a). The O_2-pulse technique is capable of two modifications. First, electron acceptors other than O_2 may be employed in order to select out limited regions of the respiratory chain, in which case it is inappropriate to refer to a H^+/O ratio and the term $H^+/2e^-$ is employed. Second, the charge stoicheiometry (q^+/O or $q^+/2e^-$) may be determined instead of the proton stoicheiometry by quantitating the movement of K^+. Even if no species other than H^+ and K^+ cross the membrane, charge- and proton-stoicheiometry are not synonymous (see Fig. 4.10).

An alternative method for determining H^+/O ratios is based on the measurement of the initial rates of respiration and proton extrusion when substrate is added to substrate-depleted mitochondria (Brand *et al.* 1976).

Any stoicheiometry determined by these methods has to satisfy the constraints imposed by thermodynamics. In other words the energy conserved in the proton electrochemical potential has to lie within the limits imposed by the redox span of the proton-translocating region. In addition, the proton-translocating regions of complexes I and III are known to be in near-equilibrium as they can be readily reversed. Therefore an approximate stoicheiometry for these regions can be deduced on purely thermodynamic grounds, knowing ΔE_h (Section 3.8) and the components of $\Delta\tilde{\mu}_{H+}$.

The importance of unequivocal $H^+/2e^-$ stoicheiometries for the proton-translocating regions of the respiratory chain is that they may be used to test mechanistic models for proton translocation. The direct group-translocation mechanism proposed by Mitchell (Section 1.4), in which protons are extruded as a consequence of the transfer of electrons from $(H^+ + e^-)$-carriers to e^--carriers requires that the observed stoicheiometry be precisely $2H^+/2e^-$ for each "loop" (Fig. 4.10). This direct mechanism also makes a number of structural demands which are discussed in Section 5.4.

Early measurements of H^+/O ratios by the O_2-pulse method (Mitchell & Moyle 1967a) gave values for $NADH \rightarrow O_2$ and succinate $\rightarrow O_2$ close to those predicted by the direct mechanism (6 and 4 respectively in Fig. 4.10). However, the finding that N-ethylmaleimide, which inhibits the H/Pi symporter (see above) increased the observed stoicheiometries started a lively controversy.

If the $H^+/2e^-$ stoicheiometries are too high to be accounted for by a direct loop mechanism, a conformational pump has to be proposed (Section 5.4). Such a proposal makes no testable predictions, in contrast to the loop

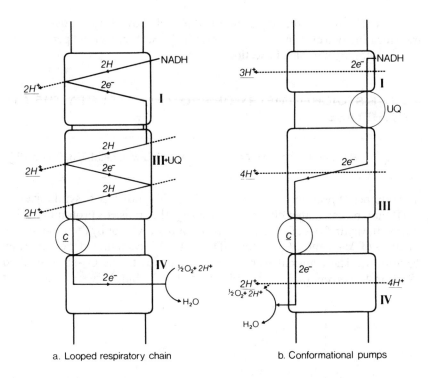

a. Looped respiratory chain   b. Conformational pumps

Fig. 4.10   Proton and charge stoicheiometry in the respiration chain.

| | Model a | | | Model b | |
	$H^+/2e^-$	$q^+/2e^-$		$H^+/2e^-$	$q^+/2e^-$
Complex I	2	2		3	3
Complex III	4	2		4	2
Complex IV	0	2		2	4
Succinate-O_2	4	4		6	6
NADH-O_2	6	6		9	9

Note that $q^+/2e^-$ ratios refer to charge movement across the membrane, and take account of both protons and electrons. Stoicheiometries for the conformational model are intended to be illustrative rather than definitive.

hypothesis, since mechanisms can be devised to account for any observed stoicheiometry. In Fig. 4.10 the relative features of the loop and conformational pump mechanisms are presented. As there is wide disagreement about the precise numbers, a middle-of-the-road set of figures is presented. It should be noted that both mechanisms predict the same stoicheiometry for complex III, but that the conformational model allows for the possibility of translocating two extra protons by complex IV (Section 5.9) and perhaps one extra proton by complex I (Section 5.6).

4.4 THE STOICHEIOMETRY OF PROTON UPTAKE BY THE ATP SYNTHETASE

Reviews Mitchell & Moyle 1968, Moyle & Mitchell 1973, Brand 1977

The number of protons translocated by the ATP synthetase to synthesize one ATP may be calculated either by measuring the transient proton extrusion during the hydrolysis of a small known amount of ATP or by thermodynamic analysis of $\Delta\tilde{\mu}_{H+}$ and ΔG_p for the ATP/ADP+Pi reaction under equilibrium conditions. The transient technique is analogous to the O_2-pulse method (Section 4.3) except that ATP is substituted for O_2, but there are a number of

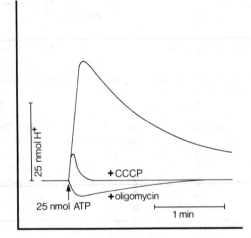

Fig. 4.11   Determination of the H^+/ATP stoicheiometry of sub-mitochondrial particles.

Sub-mitochondrial particles from beef heart were incubated in a lightly buffered KCl medium in the presence of valinomycin. No respiratory substrate was added. $2mM\ Mg^{2+}$ was present, as ATP reacts as the Mg^{2+}-complex. The initial pH of 6·1 was chosen so that no scalar protons were evolved from the hydrolysis of ATP. When the trace had stabilized, 25 nmol of MgATP was added. No acidification was seen in the presence of oligomycin, while in the presence of CCCP, a proton translocator, a brief acidification and rapid decay occurred. (After Thayer & Hinkle 1973.)

additional problems. First, the ATPase reaction itself can generate scalar protons (Eq 3.9). Secondly, entry of ATP and exit of ADP and Pi generates vectorial protons (Section 7.5). The second problem may be overcome by working with inverted sub-mitochondrial particles (Section 1.4), and the first by adjusting the pH so that there are no scalar protons (Thayer & Hinkle 1973). A value close to $2H^+/ATP$ is obtained for the mitochondrial ATP synthetase (Fig. 4.11).

The thermodynamic approach has been used for both mitochondria, sub-mitochondrial particles, bacterial vesicles and chloroplasts. In the case of intact mitochondria, a value of 3 is frequently, but not invariably, found by this approach (Fig. 4.12). Two of the protons can be ascribed to the synthetase itself, and the third to the transport of ADP, Pi and ATP (Section 7.5).

The stoicheiometries for proton extrusion by the respiratory chain and for

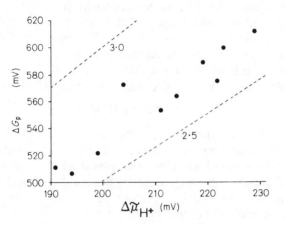

Fig. 4.12 Thermodynamic relationship between the extra-mitochondrial phosphorylation potential and $\Delta\tilde{\mu}_{H+}$.
Brown adipose tissue mitochondria were incubated in a medium containing α-glycerol-3-phosphate as substrate, GDP to inhibit the proton short-circuit operative in these mitochondria (Section 4.5), valinomycin, ^{86}Rb, ^{14}C-methylamine and ^{3}H-acetate to enable $\Delta\tilde{\mu}_{H+}$ to be measured (Section 4.2). ADP was present, together with increasing, sub-optimal additions of the proton translocator FCCP (Section 2.5). $\Delta\tilde{\mu}_{H+}$ was measured for each FCCP concentration once steady-states had been attained. In a parallel experiment, the same FCCP concentrations were added to an identical incubation which lacked the isotopes but contained ^{3}H-ADP. When steady-states were attained, the incubation was analysed for ^{3}H-ADP and ^{3}H-ATP to enable the phosphorylation potential (expressed here in mV) to be calculated. The diagonal dotted lines represent the highest phosphorylation potentials which could be maintained for a given $\Delta\tilde{\mu}_{H+}$ if either 2·5 or 3 protons were used in the synthesis and transport of 1 ATP. (Data from Nicholls & Bernson 1977.)

proton re-entry during ATP synthesis have to be consistent with the overall observed stoicheiometry for ATP synthesis (Section 4.6). Thus:

$$ADP/2e^- = H^+/2e^- \div H^+/ATP \qquad \text{(Eq. 4.1)}$$

The generally accepted maximal value for $ADP/2e^-$ of 3 for $NADH \rightarrow O_2$ and 2 for succinate$\rightarrow O_2$ (Section 4.6), combined with a H^+/ATP ratio of 3 for ATP synthesis and export, would suggest a H^+/O ratio of 9 and 6 respectively, which would favour a conformational pump model rather than a looped respiratory chain, as the latter is restricted to values of 6 and 4 respectively (Fig. 4.10).

4.5 PROTON CURRENT, PROTON CONDUCTANCE AND RESPIRATORY CONTROL

The previous sections have been concerned with the potential term in the proton circuit and with the gearing of the transducing complexes for the generation and utilization of this potential. This section will deal with the factors which regulate the proton current in the circuit.

The current of protons flowing around the proton circuit (J_{H+}) may be readily calculated from the rate of respiration and the H^+/O stoicheiometry:

$$J_{H+} = dO/dt \times H^+/O \qquad \text{(Eq. 4.2)}$$

For a given substrate therefore, proton current and respiratory rate vary in parallel. The most important aspects of the control of mitochondrial respiration were established before the advent of the chemiosmotic hypothesis, and one of the theory's successes lies in the straightforward way in which it is possible to explain the action of a large number of agents in terms of their modulation of the proton current.

Since the 1950s the oxygen electrode (Fig. 4.13) has proved an extremely versatile tool for investigating mitochondria. The oxygen electrode only determines directly the rate of a single reaction, i.e. the transfer of electrons to O_2, and to obtain information on other mitochondrial processes it is necessary to arrange the incubation conditions so that the desired process becomes rate-limiting. Possible rate-limiting steps which may be devised include (Fig. 4.14):

(a) substrate transport across the membrane
(b) substrate dehydrogenase activity
(c) respiratory chain activity
(d) adenine nucleotide transport across the membrane
(e) ATP synthetase activity
(f) H^+-permeability of the membrane

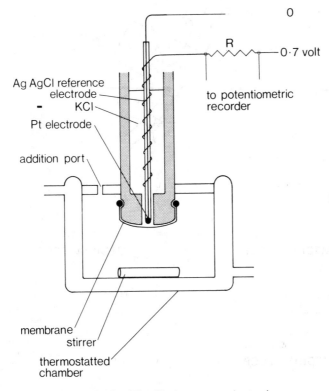

Fig. 4.13   The Clark oxygen electrode

At the Pt electrode, O_2 is reduced to H_2O. If the Pt electrode is maintained about 0·7 V negative with respect to the Ag/AgCl reference, a current will flow which is proportional to the rate of O_2 consumption by the electrode. Under appropriate conditions, the current is proportional to the O_2 concentration in the solution. A thin O_2-permeable membrane prevents the electrode from becoming poisoned. Because of the oxygen consumption by the electrode, the incubation must be stirred continuously to prevent a depletion layer forming at the membrane. The chamber is sealed except for a small addition port. The electrode is calibrated with air-saturated medium, and under anoxic conditions following dithionite addition. A typical chamber has a volume of 2 ml, and requires 2–3 mg of mitochondrial protein.

Three basic states of the proton circuit were shown in Fig. 4.1: open circuit, where there is no evident means of proton re-entry into the matrix; a circuit completed by proton re-entry coupled to ATP synthesis; and a circuit completed by a proton leak not coupled to ATP synthesis. These states can readily be created in the O_2-electrode chamber (Fig. 4.15) and are referred to using the convention proposed by Chance & Williams (1955) (Table 4.1).

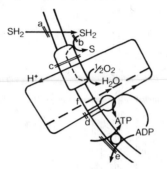

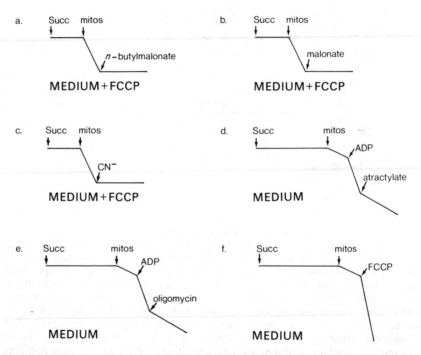

Fig. 4.14 The use of the oxygen electrode in mitochondrial energy transduction. In the scheme, six possible ways of interfering with mitochondrial energy transduction are shown. The diagramatic oxygen electrode traces show how these perturbations might be investigated with the oxygen electrode. The incubation medium is presumed to contain osmotic support, pH buffering and Pi.

- (*a*) substrate transport inhibited
- (*b*) substrate dehydrogenase inhibited
- (*c*) respiratory chain inhibited
- (*d*) adenine nucleotide translocator inhibited
- (*e*) ATP synthetase inhibited
- (*f*) proton translocator added

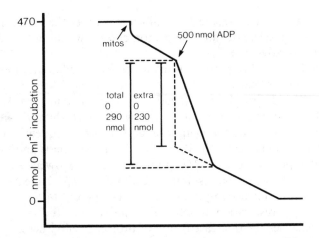

Fig. 4.15 The determination of ADP/O ratios using the oxygen electrode. In this example, rat liver mitochondria were incubated in the presence of succinate as substrate (rotenone being present to inhibit any endogenous NAD^+-linked respiration). Where shown, ADP was added. When virtually all the ADP had been phosphorylated, respiration returned to the low State 4 rate. If it is assumed that the proton leakage giving rise to the State 4 respiration continues at the same rate during accelerated State 3 respiration, it is appropriate to take only the *extra* O in the calculation of the ADP/O ratio. If, on the other hand, it is assumed that the leakage stops during State 3, then all the respiration is being utilized for ATP synthesis and the total O is used. In practice the latter convention is more accurate, as proton leakage is highly dependent on $\Delta\tilde{\mu}_{H^+}$, and the slight decrease (Fig. 4.16) largely abolishes it. Thus ADP/O in this example=500/290=1·72.

The addition of mitochondria to the incubation (Fig. 4.17) causes an initial burst of respiration which then subsides to a low, State 4 rate. Although mitochondria contain adenine nucleotides within their matrices, the amount is relatively small (about 10 nmol mg protein^{-1}), and when the mitochondria are introduced into the incubation this pool will be very rapidly phosphorylated until it achieves equilibrium with $\Delta\tilde{\mu}_{H^+}$. The subsequent State 4 respiration occurs because the inner membrane is not completely impermeable to protons, which can slowly leak back across the membrane even in the absence of net ATP synthesis. One factor which contributes to this State 4 proton leak is the slow cycling of Ca^{2+} across the membrane (Section 8.4).

The factor which actually controls the rate of respiration is the extent of disequilibrium between the redox potential spans across the proton-translocating regions of the respiratory chain and $\Delta\tilde{\mu}_{H^+}$ (Section 3.9). In State 4, respiration is automatically regulated so that the rate of proton extrusion by the respiratory chain precisely balances the rate of proton leak back across the membrane. If proton extrusion were momentarily to exceed the rate of re-entry, $\Delta\tilde{\mu}_{H+}$ would increase, the disequilibrium between the respiratory chain and $\Delta\tilde{\mu}_{H+}$ would decrease, and the proton current would in turn decrease, restoring the steady-state.

In Fig. 4.15 State 4 respiration is disturbed by the addition of exogenous ADP. This allows matrix ATP to exchange for the added ADP via the adenine nucleotide translocator (Section 7.5). As a result, the ΔG for the ATP/ADP+Pi system in the matrix is lowered, disturbing the ATP synthetase equilibrium. The following events then occur sequentially: (a) the ATP synthetase operates in the direction of ATP synthesis and proton re-entry to attempt to restore the ΔG; (b) proton re-entry lowers $\Delta\tilde{\mu}_{H+}$; (c) the disequilibrium between the respiratory claim and $\Delta\tilde{\mu}_{H+}$ increases; (d) the proton current and hence respiration increases. As for State 4, this accelerated State 3_{ADP} respiration is self-regulating so that the rate of proton extrusion balances the (increased) rate of proton re-entry across the membrane. Net ATP synthesis, and State 3_{ADP} respiration, may be terminated in three ways: (a) when sufficient ADP is phosphorylated to ATP to regain thermodynamic equilibrium; (b) by inhibiting adenine nucleotide exchange across the membrane by adding an inhibitor such as atractylate (Section 7.6); (c) by inhibiting the ATP synthetase, for example by the addition of oligomycin (Section 7.3).

Energy transduction between the respiratory chain and the proton electrochemical potential is extremely efficient, in that a small thermodynamic disequilibrium between the two can result in a considerable energy flux. Thus Fig. 4.16 shows that $\Delta\tilde{\mu}_{H+}$ drops by less than 30% when ADP is added to induce rapid State 3_{ADP} respiration. The actual disequilibrium between the respiratory chain and $\Delta\tilde{\mu}_{H+}$ is even less, as the redox spans also decrease (Section 5.3).

Efficient energy-transduction during State 3_{ADP} is also apparent at the ATP synthetase. A high rate of ATP synthesis can be maintained with only a slight thermodynamic disequilibrium between $\Delta\tilde{\mu}_{H+}$ and ΔG_p.

Proton translocators uncouple oxidative phosphorylation by inducing an artificial proton permeability in bilayer regions of the membranes (Section 2.5). They may thus be used to over-ride the inhibition of proton re-entry which results from an inhibition of net ATP synthesis. As a consequence proton translocators such as FCCP can induce rapid State 3_{unc} respiration, regardless of the presence of oligomycin or atractylate, or the absence of ADP (Fig. 4.14).

The respiratory chain responds to a $\Delta\tilde{\mu}_{H+}$ which is lowered by the addition of a proton translocator in the same way as to a $\Delta\tilde{\mu}_{H+}$ lowered by ATP synthesis. In both cases the rate of proton extrusion (and respiration) adjusts until the rate exactly balances the rate of proton re-entry across the inner membrane. The resulting $\Delta\tilde{\mu}_{H+}$ stabilizes when the rate at which $\Delta\tilde{\mu}_{H+}$ drives

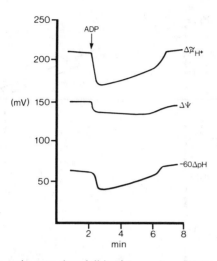

Fig. 4.16   There is only a modest fall in $\Delta\tilde{\mu}_{H+}$ when ADP is added to induce a
State 4–State 3 transition in liver mitochondria.
Rat liver mitochondria were incubated in a sucrose-based medium containing
B-hydroxybutyrate as substrate, $^{86}Rb^+$ and valinomycin to measure $\Delta\Psi$, and
^{14}C-methylamine and ^3H-acetate to measure ΔpH. At the point indicated
(*arrow*) sufficient ADP was added to sustain State 3 respiration for 5 min. $\Delta\Psi$
and ΔpH were measured by membrane filtration (from Nicholls 1974).

proton re-entry is matched by the rate at which the disequilibrium between
the respiratory chain and $\Delta\tilde{\mu}_{H+}$ drives proton extrusion. Respiration does not
continue to increase indefinitely as more proton translocator is added to the
incubation, because a stage is reached when kinetic factors in the respiratory
chain become rate limiting. Respiration is now *uncontrolled*, i.e. it is no
longer dependent on the thermodynamic disequilibrium. Uncontrolled res-
piration generally sets in when $\Delta\tilde{\mu}_{H+}$ is still very high (Fig. 4.17). It is essential
to distinguish between the condition where no respiratory control is evident
($\Delta\tilde{\mu}_{H+}$ = 0 to 170 mV; Fig. 4.17) and the situation where mitochondria are
strictly de-energized ($\Delta\tilde{\mu}_{H+}$ = 0).

In an electrical circuit, the conductance of a component is calculated from
the current flowing per unit potential difference. A similar calculation for the
proton circuit enables the effective proton conductance of the membrane
(C_MH^+) to be calculated:

p. 89   Equation 4.3 should read
$$C_MH^+ = J_{H^+}/\Delta\tilde{\mu}_{H^+}$$
(4.3)

In Table 4.3 the calculation of C_MH^+ for a typical preparation of liver mitochondria is shown, together with other pre-chemiosmotic parameters to be
discussed in Section 4.6.

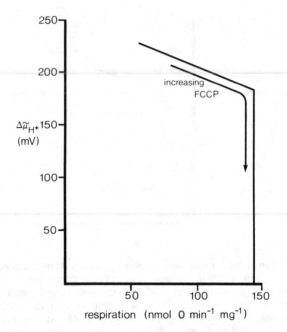

respiration (nmol 0 min^{-1} mg^{-1})

Fig. 4.17   Respiratory rate as a function of $\Delta\tilde{\mu}_{H+}$.
Mitochondria from brown adipose tissue were incubated in a medium containing
glycerol-3-phosphate as substrate, oligomycin to inhibit ATP synthetase, and
GDP to inhibit the native proton-conductor peculiar to these mitochondria
(Section 4.5). Increasing concentrations of proton translocator were added, and
respiration and $\Delta\tilde{\mu}_{H+}$ were determined in parallel (from Nicholls & Bernson 1977).

The magnitude of the endogenous proton conductance of the membrane is
the parameter which underlies the bioenergetic behaviour of a given preparation of mitochondria. As will be shown in Section 4.6 the overall, non-
chemiosmotic parameters of "energization" may each be referred back to
$C_M H^+$. Evidently, for an efficient transduction of energy, $C_M H^+$ should be
as low as possible. There is, however, one class of mitochondria where there
appears to be a physiological increase in the proton conductance of the inner
membrane. Brown adipose tissue is capable of very rapid respiration, either
for thermogenesis or to dissipate excess substrates as a means of obesity
regulation. In this tissue the normal control of respiration by ATP demand is
over-ridden by a protein in the inner membrane which acts as a regulatable
proton translocator, allowing proton re-entry, and hence State 3 respiration,
to occur without the necessity for stoicheiometric ATP synthesis (Nicholls
1976, 1979). The activity of the protein is sufficient to increase $C_M H^+$ to a
value 30-fold greater than that characteristic of other mitochondria.

Table 4.3 The calculation of parameters of mitochondrial energy transduction

Problem A typical preparation of liver mitochondria oxidizing succinate in State 4 (Table 4.1) has a respiration rate of 15 nmol of O min^{-1} mg protein^{-1} and sustains a $\Delta\tilde{\mu}_{H^+}$ of 220 mV. Addition of a low concentration of the proton translocator FCCP increases respiration to 100 nmol of O min^{-1} mg protein^{-1} and lowers $\Delta\tilde{\mu}_{H^+}$ to 40 mV. If the H$^+$/O ratio for succinate to O$_2$ is assumed to be 6, calculate the respiratory control ratio and the proton current and effective proton conductance before and after FCCP addition.

(a) Respiratory control ratio = $\dfrac{\text{respiration after FCCP}}{\text{respiration before FCCP}}$ =100/6=6·7

(b) Proton current=respiration × H$^+$/O ratio= 90 (−FCCP)
600 (+FCCP)
(units nmol H$^+$ min^{-1} mg protein^{-1})

Proton conductance $C_M H^+$ = $\dfrac{\text{proton current}}{\Delta\tilde{\mu}_{H^+}}$ = 90/220=0·41 (−FCCP)
600/40=15 (+FCCP)
(units nmol H$^+$ min^{-1} mg protein^{-1} mV^{-1})

4.6 NON-CHEMIOSMOTIC PARAMETERS OF ENERGY-TRANSDUCTION

4.6.1 Respiratory control ratio

This is a pre-chemiosmotic parameter for assessing the integrity of a mitochondrial preparation. It is defined as the respiratory rate attained during maximal ATP synthesis (i.e. in the presence of ADP), or in the presence of a proton translocator, divided by the rate in the absence of ATP synthesis or proton translocator. Thus:

$$\text{Respiratory control ratio} = \frac{\text{State 3 respiration}}{\text{State 4 respiration}} \quad \text{(Eq. 4.4)}$$

Typical values for the ratio vary from 3 to 15 in different preparations. Although this parameter is useful empirically, it is important to realize that it is a hybrid function: the State 4 rate depends on the native $C_M H^+$ of the membrane, while the State 3 rate is dependent upon whichever step is rate limiting in State 3, be it substrate permeability, substrate dehydrogenase activity, respiratory chain, ATP synthesis or adenine nucleotide translocation. Respiratory control ratios must therefore be interpreted with caution. The State 4 rate itself would be a better criterion of the tightness of coupling between respiration and ATP synthesis.

4.6.2 ADP/O and P/O ratios

While the stoicheiometries of proton translocation by the respiratory chain and ATP synthetase are fixed, even if the actual values are still a matter for contention, the overall stoicheiometry of ATP synthesis in relation to respiration can vary from a theoretical maximum of about one ATP synthesized per $2e^-$ per energy-transducing region down to zero, depending on the activity of the parallel proton leak pathway bypassing the ATP synthetase (Fig. 4.1). Any factor which increases the leak conductance will decrease the proportion of protons which re-enter through the ATP synthetase. A second factor is that as $\Delta\tilde{\mu}_{H^+}$ decreases when $C_M H^+$ increases (Fig. 4.17), the thermodynamic ability to synthesize ATP decreases. Because it is difficult to estimate the leakage current during ATP synthesis, there are considerable errors in calculated values, as these are usually calculated for the extrapolated case of zero proton leak.

4.6.2.1 ADP/O ratios by the oxygen electrode technique

This method determines the extent of the burst of accelerated State 3 respiration obtained when a small measured aliquot of ADP is added to mitochondria respiring in State 4, Fig. 4.17. Almost all the added ADP is phosphorylated to ATP, the ATP:ADP ratio being typically 100:1 when State 4 is regained, and the ratio *moles of ADP added/moles of O consumed* can be calculated. In order to correct for the proton leak, it is the convention to assume that the leak ceases during State 3 respiration. On first sight it might be thought that as $\Delta\tilde{\mu}_{H^+}$ falls only by about 30% during the ADP cycle (Fig. 4.16) that the leak would continue unabated. However, in practice the leak conductance shuts off almost completely when $\Delta\tilde{\mu}_{H^+}$ drops below 200 mV, so that it is more correct to consider the total O uptake during the cycle in the calculation of the ratio.

4.6.2.2 P/O ratios by the glucose-hexokinase trap

The second method relies on the synthesis of $[\gamma\text{-}^{32}P]$-ATP from ADP and ^{32}Pi, and the subsequent trapping of the label, and regeneration of ADP, by means of added hexokinase:

$$\text{ADP} + {}^{32}\text{Pi} \rightarrow [\gamma\text{-}^{32}\text{P}]\text{-ATP}$$

$$[\gamma\text{-}^{32}\text{P}]\text{-ATP} + \text{glucose} \rightarrow \text{glucose-6-}^{32}\text{P} + \text{ADP}$$

For this method, respiration is monitored with an O_2 electrode: after a defined oxygen uptake, the reactions are quenched, and the glucose-6-^{32}P is separated from the ^{32}Pi and counted. No cycles of respiratory stimulation are observed as ADP is continuously regenerated. In contrast to the O_2

electrode technique, the ΔG_p is maintained at a low value by the hexokinase. ATP synthesis can therefore be detected under quite unfavourable conditions.

Values for $ADP/2e^-$ ratios are, as with all stoicheiometries, a source of debate. The "classic" value of 1 ATP per $2e^-$ per proton-translocating complex is only consistent with H^+/O ratios of 6 and 4 for $NADH-O_2$ and succinate–O_2 respectively if only two protons are used for the synthesis of ATP and for the transport of adenine nucleotides and Pi across the inner membrane. However, there is strong evidence that the transport consumes a third proton (Section 7.6). The higher H^+/O ratios of 9 and 6 (Fig. 4.10) for NADH and succinate are consistent with a H^+/ATP of 3 including transport. Brand et $al.$ (1978) have suggested that the $ADP/2e^-$ ratios for succinate to cyt c, and for cyt c to O_2 are non-integral, 0·67 and 1·33 respectively.

4.7 REVERSED ELECTRON TRANSFER AND THE PROTON CIRCUIT DRIVEN BY ATP HYDROLYSIS

The ATP synthetase is reversible and is only constrained to run in the direction of net ATP synthesis by the continual regeneration of $\Delta\tilde{\mu}_{H+}$ and the use of ATP by the cell. If therefore the respiratory chain is inhibited and ATP is supplied to the mitochondrion, the ATP synthetase functions as an ATPase, generating a $\Delta\tilde{\mu}_{H+}$ comparable to that produced by the respiratory chain (Fig. 4.7). The proton circuit generated by ATP hydrolysis must be completed by a means of proton re-entry into the matrix. Proton translocators therefore accelerate the rate of ATP hydrolysis, just as they accelerate the rate of respiration; this is the "uncoupler-stimulated ATPase activity".

The classic means of discriminating whether a mitochondrial energy-dependent process is driven directly by $\Delta\tilde{\mu}_{H+}$, or indirectly via ATP, is to investigate the sensitivity of the process to the ATP synthetase inhibitor oligomycin (Section 1.4). A $\Delta\tilde{\mu}_{H+}$-driven event would be insensitive to oligomycin when the potential was generated by respiration, but sensitive when $\Delta\tilde{\mu}_{H+}$ was produced by ATP hydrolysis. The converse would be true of an ATP-dependent event (Fig. 4.18).

The near equilibrium in State 4 between $\Delta\tilde{\mu}_{H+}$ and the redox spans of complexes I and III (Fig. 4.2) suggests that conditions could be devised in which these segments of the respiratory chain could be induced to run backwards, driven by the inward flux of protons. It should be noted that this does not apply to complex IV, which is essentially irreversible. Reversed electron transfer may be induced in two ways, either by generating a $\Delta\tilde{\mu}_{H+}$ by ATP hydrolysis, or by using the flow of electrons from succinate or cytochrome c to O_2 to reverse electron transfer through Complexes I or I and III respectively (Fig. 4.18).

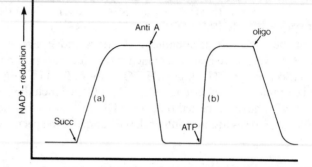

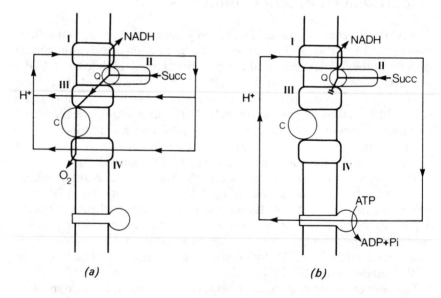

Fig. 4.18 Reversed electron transfer in the mitochondrial respiratory chain. Schematic response of sonicated sub-mitochondrial particles incubated in the presence of NAD$^+$. In (a) NAD$^+$ reduction (measured at 340 nm) is induced by reversed electron transfer in complex I. Succinate acts both as donor of electrons for reversed electron transfer and as substrate for complexes III and IV. In (b) succinate merely donates electrons to NAD$^+$, $\Delta\bar{\mu}_{H^+}$ is generated by ATP hydrolysis, complex III being inhibited by antimycin A.

Under physiological conditions the mitochondrial ATP synthetase will not normally be called upon to act as a proton-translocating ATPase, except possibly during periods of anoxia when glycolytic ATP could be utilized to maintain the mitochondrial $\Delta\tilde{\mu}_{H+}$. However, some bacteria, such as *Streptococcus faecalis* when grown on glucose, lack a functional respiratory chain and rely entirely upon hydrolysis of glycolytic ATP to generate a $\Delta\tilde{\mu}_{H+}$ across their membrane and enable them to transport metabolites (Harold 1977).

4.8 ATP SYNTHESIS DRIVEN BY AN ARTIFICIAL PROTON ELECTROCHEMICAL POTENTIAL

Reviews Jagendorf 1975, Mitchell 1976b, Kagawa 1978, Fillingame 1980

The chemiosmotic hypothesis predicts that an artificial generated $\Delta\tilde{\mu}_{H+}$ should be able to cause the net synthesis of ATP in any energy-transducing membrane with a functional ATP synthetase. The first demonstration that this was so came from chloroplasts (Jagendorf & Uribe 1966). These authors

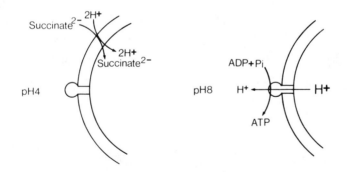

Fig. 4.19 The "acid-bath" experiment—a ΔpH can generate ATP Chloroplasts with broken envelopes (Section 1.3) were incubated in the dark at pH 4 in a medium containing succinate. Inhibitors were present to inhibit any electron transfer. The succinate slowly pereated into the thylakoid space, liberating protons and lowering the thylakoid space pH to about 4 (*left*). The external pH was then suddenly raised to 8, creating a ΔpH of 4 units across the membrane, and simultaneously ADP, Pi and Mg^{2+} were added (*right*). Proton efflux through the ATP synthetase led to the synthesis of up to 100 moles of ATP per mole of synthetase. Synthetic proton translocators such as FCCP inhibited the ATP production. (Data from Jagendorf & Uribe 1966.)

found that chloroplasts equilibrated in the dark at acid pH could be induced to synthesize ATP when the external pH was suddenly increased from 4 to 8, creating a transitory pH gradient of 4 units across the membrane (Fig. 4.19).

Chloroplasts normally operate with a high ΔpH and a low $\Delta\Psi$. With mitochondria the opposite is the case, and the first successful demonstration of net ATP synthesis by an artificial $\Delta\tilde{\mu}_{H+}$ occurred when both a pH gradient and a K^+ diffusion potential where created across the membrane of sub-mitochondrial particles (Thayer & Hinkle 1975).

Experiments with intact organelles show that ATP can be synthesized with $\Delta\tilde{\mu}_{H+}$ as sole energy input, but they do not distinguish between models of energy transduction in which $\Delta\tilde{\mu}_{H+}$ is the sole link between electron transfer and ATP synthesis, and those in which a proton pump is on a side-path in equilibrium with an alternative intermediate (Fig. 1.11). For this it is necessary to purify the ATP synthetase until it possesses no component in common with proton-translocating preparations of the respiratory chain, to reconstitute it into a synthetic membrane (Section 1.3), and then to apply an artificial $\Delta\tilde{\mu}_{H+}$. This has been done with a highly purified ATP synthetase from thermophilic bacteria (Sone et al. 1977).

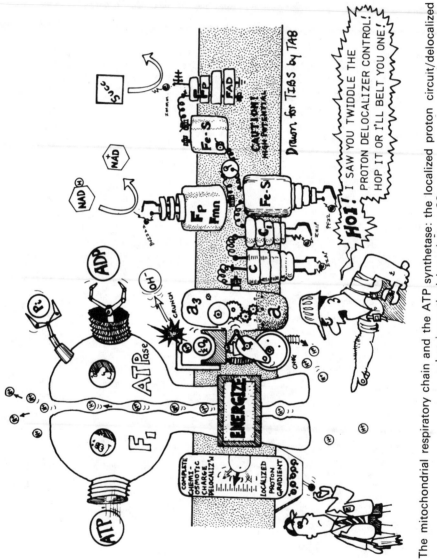

The mitochondrial respiratory chain and the ATP synthetase: the localized proton circuit/delocalized chemiosmotic debate (see p. 22)

Respiratory Chains

5.1 INTRODUCTION

The respiratory chain of mitochondria is an assembly of more than twenty discrete carriers of electrons, together with an unspecified number of "structural" peptides (Fig. 5.1). While the function of the respiratory chain as an oxidation–reduction driven proton pump is now accepted, the structural basis underlying this function is only partially understood. This chapter will look at the approaches which have been taken to investigate the structure of the respiratory chains of mitochondria and bacteria.

5.2 COMPONENTS OF THE MITOCHONDRIAL RESPIRATORY CHAIN, AND METHODS OF DETECTION

Review Slater 1974

The respiratory chain transfers electrons through a redox potential span of 1·1 volts, from the $NAD^+/NADH$ couple to the $O_2/2H_2O$ couple. Much of the respiratory chain is reversible (Section 4.7), and to catalyse both the forward and reverse reactions it is necessary for the redox components to operate under conditions where both the oxidized and reduced forms exist at appreciable concentrations. In other words, the operating redox potential of a couple, E_h, (Section 3.3) should not be far removed from the mid-point potential of the couple, E_m. As will be shown in Fig. 5.11, this constraint is generally obeyed, and this in turn gives some rationale to the apparently random selection of redox carriers within the respiratory chain.

The initial transfer of electrons from the soluble dehydrogenases of the citric acid cycle requires a cofactor which has a mid-point potential in the region of $-300\,mV$ and is sufficiently mobile to shuttle between the matrix dehydrogenases and the membrane-bound respiratory chain. This function is filled by the $NADH/NAD^+$ couple, which has an $E_{m,7}$ of $-320\,mV$ (Fig. 5.2).

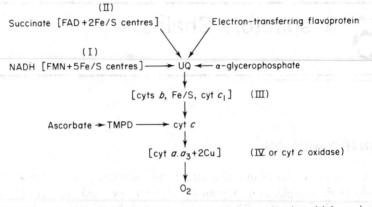

Fig. 5.1 The linear sequence of redox carriers in the mitochondrial respiratory chain.
The groups of redox carriers within the brackets represent the structural complexes I to IV.

While the majority of electrons are transferred to the respiratory chain in this way, a group of enzymes catalyse dehydrogenations where the mid-point potential of the substrate couple is close to 0 mV. These, succinate dehydrogenase, α-glycerophosphate dehydrogenase, and the "electron-transferring flavoprotein" (transferring electrons from fatty acid β-oxidation), feed electrons directly into the respiratory chain at a potential close to 0 mV without the intermediacy of the $NAD^+/NADH$ couple (Fig. 5.1). This direct transfer requires that these enzymes be in direct contact with the respiratory chain, in other words membrane-bound.

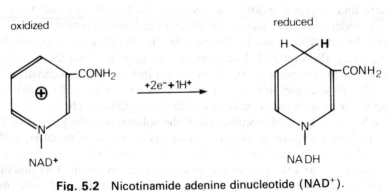

Fig. 5.2 Nicotinamide adenine dinucleotide (NAD^+).
NAD^+ Nicotinamide-ribose-P-O-P-ribose-adenine.

7,8 - Dimethylisoalloxazine

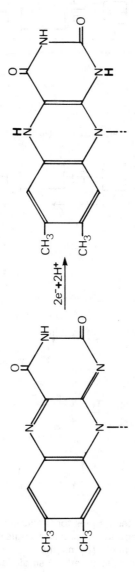

oxidized

reduced

Fig. 5.3   Flavin mononucleotide (FMN) and flavin adenine dinucleotide (FAD).

FMN   7,8-Dimethylisoalloxazine-D-ribitol-P
       └─────────────┬─────────────┘
                 riboflavin

FAD   7,8-Dimethylisoalloxazine-D-ribitol-P-O-P-ribose-adenine

a. Split-beam spectrophotometer

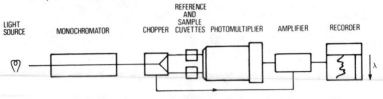

LIGHT SOURCE   MONOCHROMATOR   CHOPPER   REFERENCE AND SAMPLE CUVETTES   PHOTOMULTIPLIER   AMPLIFIER   RECORDER

b. Dual-wavelength spectrophotometer

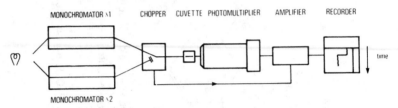

MONOCHROMATOR λ1   CHOPPER   CUVETTE   PHOTOMULTIPLIER   AMPLIFIER   RECORDER

MONOCHROMATOR λ2

c. Rapid kinetics: stopped flow in combination with dual wavelength

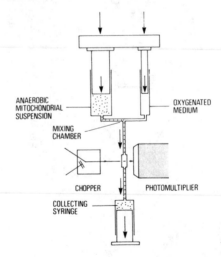

ANAEROBIC MITOCHONDRIAL SUSPENSION

OXYGENATED MEDIUM

MIXING CHAMBER

CHOPPER   PHOTOMULTIPLIER

COLLECTING SYRINGE

Fig. 5.4 Spectroscopic techniques for the study of the respiratory chain. (*a*) The split-beam spectrophotometer uses a single monochromator, the output from which is directed alternately (by means of a chopper oscillating at about 300 Hz) into reference and sample cuvette. A single large photomultiplier is used, and the alternating signal is amplified and decoded such that the output from the amplifier is proportional to the difference in absorption between the two cuvettes. If the monochromator wavelength is steadily increased, a (sample) − (reference) difference spectrum will be obtained. The split beam is therefore used to plot difference spectra which do not change with time.

The redox carriers within the respiratory chain consist of flavoproteins, which contain tightly bound FAD or FMN as prosthetic groups (Fig. 5.3) and undergo a $(2H^+ + 2e^-)$ reduction; cytochromes, with porphyrin prosthetic groups undergoing a $1e^-$ reduction; iron–sulphur (or non-haem iron) proteins which possess prosthetic groups also reduced in a $1e^-$ step; ubiquinone, which is a free, lipid-soluble cofactor reduced by $(2H^+ + 2e^-)$; and finally protein-bound Cu, reducable from Cu^{II} to Cu^I.

Cytochromes are classified according to the structure of their porphyrin prosthetic group. Mitochondria contain a, b and c-type cytochromes. Cytochromes d and o occur in some bacterial chains as terminal oxidases, o cytochromes being autoxidizable b-type cytochromes, while d-cytochromes possess a partially saturated chlorin ring in place of porphyrin.

The cytochromes were the first components to be detected, due to their distinctive visible spectra. An individual cytochrome exhibits one major absorption band in its oxidized form, while most cytochromes show three absorption bands when reduced. Absolute spectra, however, are of limited use when studying cytochromes in intact mitochondria, owing to the high non-specific absorption and light-scattering of the organelles, particularly as the light-scattering can change as a consequence of metabolism-induced matrix volume changes (Section 2.7). For this reason, cytochrome spectra are studied using a sensitive differential, or split-beam, spectroscopy in which light of a steadily increasing wavelength is split between two cuvettes containing incubations of mitochondria identical in all respects except that an addition is made to one cuvette to create a differential reduction of the cytochromes (Fig. 5.4). The output from the reference cuvette is then automatically substracted from that of the sample cuvette, to subtract all non-specific absorption. Figure 5.5 shows the reduced, oxidized, and reduced

---

(b) The dual wavelength spectrophotometer uses two monochromators: one set at a wavelength which is optimal for the transient in question and one set for a near-by isobestic point at which no change occurs. Light from the two monochromators is sent alternately through a single cuvette while the transient is induced. The amplifier decodes the photomultiplier output and the recorder plots the difference in extinction $(\lambda_1 - \lambda_2)$ during the transient against time. The dual wavelength therefore is used to follow the kinetics of a given spectral component.

(c) In many cases manual mixing is too slow to allow kinetics to be resolved. The dual-wavelength can therefore be modified to allow the mitochondrial suspension to be mixed with the agent inducing the transient and introduced into the cuvette within a few milliseconds, by driving the contents of two syringes through a mixing chamber into the cuvette. While the flow continues, the "age" of the mixture will remain constant, depending on the length of tubing between the mixing chamber and the cuvette; when the flow is stopped, the mixture will "age" and this can be followed.

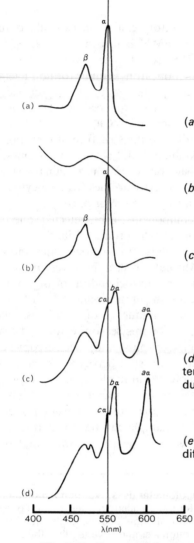

(a) Cyt c: absolute reduced
spectrum

(b) Cyt c: absolute oxidized
spectrum

(c) Cyt c: differential reduced
minus oxidized spectrum

(d) Beef heart SMPs: room
temperature differential reduced minus oxidized spectrum

(e) Beef heart SMPs: 77°K
differential reduced minus
oxidized spectrum

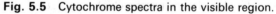

Fig. 5.5 Cytochrome spectra in the visible region.

The absolute reduced and oxidized spectra (a) and (b) were obtained with the purified cytochrome in a split-beam spectrophotometer with water in the reference cuvette. The reduced minus oxidized difference spectrum of purified cyt c (c) was obtained with reduced c in the sample cuvette and oxidized c in the reference cuvetted. (d) shows the reduced minus oxidized spectrum from beef heart sub-mitochondrial particles. Those in the sample cuvette were maintained reduced by the addition of dithionite, while ferricyanide was added to the particles in the reference cuvette. In (e) the scan was repeated at the temperature of liquid N₂ (77°K); note the greater sharpness of the α-bands. Spectra courtesy of W. J. Ingledew.

minus oxidized spectra for isolated cyt c, together with the complex reduced minus oxidized difference spectra obtained with sub-mitochondrial particles, in which the peaks of all the cytochromes are superimposed.

The individual cytochrome may most readily be resolved on the basis of their a-absorption bands in the 550–610 nm region. The sharpness of the spectral bands can be enhanced by running spectra at liquid N_2 temperatures (77°K), due to a decrease in line broadening due to molecular motion and to an increased effective light path through the sample resulting from multiple internal reflections from the ice crystals (Fig. 5.5).

Room-temperature difference spectroscopy clearly distinguishes only a single a-, b- and c-type cytochrome. However, each is now known to comprise two spectrally distinct components. The a-type cytochromes can be resolved into a and a_3 in the presence of CO, which combines specifically with a_3, but it is not clear whether a and a_3 represent chemically distinct species. The b-cytochromes consist of two components which respond differently when a $\Delta\tilde{\mu}_{H+}$ is established across the membrane (Section 5.8). The two components were labelled b_T (or b_{562}) and b_K (or b_{566}). Recent evidence (Section 5.8) suggests that the b cytochromes are a single chemical species, but that they function as dimers in the respiratory chain.

The two c-type cytochromes, cyt c and cyt c_1, can be resolved spectrally at low temperatures. Cyt c_1 is an integral protein within complex III (Section 5.8), while cyt c is a peripheral protein on the C-face of the membrane and links complex III with cytochrome c oxidase.

While their distinctive visible spectra aided the early identification and investigation of the cytochromes, the other major class of electron carriers,

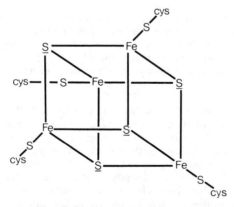

Fig. 5.6 Iron–sulphur centres.
A centre with 4Fe and 4 acid-labile sulphurs is shown. On treatment with acid these sulphurs (*underlined*) are liberated as H_2S. Although there are four Fe atoms, the entire centre undergoes only a $1e^-$ oxido-reduction.

a. Apparatus

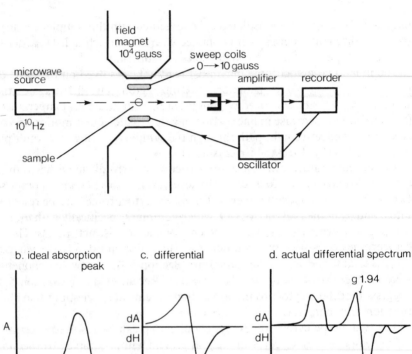

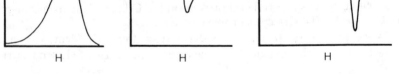

Fig. 5.7 Electron spin resonance and the detection of Fe/S centres.
(a) Apparatus: a microwave source produces monochromatic radiation at about 10^9Hz, 30 cm. Unpaired electrons in the sample absorb the radiation when a magnetic field is applied, the precise value of the field required for absorption depending on the molecular environment of the electron, according to the formula:

$$hv = g\beta H$$

where h is Planck's constant, v the frequency of the radiation, β a constant, the Bohr magneton, H the applied magnetic field, and g the spectroscopic constant which is diagnostic of the species. The spectrum is obtained by keeping the microwave frequency constant and varying the magnetic field, trace (b). In practice, a differential spectrum, trace (c), is obtained by superimposing upon the steadily increasing field a very rapid modulation of small amplitude obtained with auxilliary sweep coils. The change in microwave absorption across each of these sweeps enables the differential to be obtained. Spectra of energy-transducing membranes are complex, trace (d), g values are obtained from either the peaks, the troughs or the points of inflexion of the trace. Samples must be frozen and generally present at a high protein concentration (10–50 mg protein ml^{-1}).

the iron–sulphur (Fe/S) proteins (Fig. 5.6) have ill-defined visible spectra but characteristic electron spin resonance spectra (ESR or EPR); see Beinert (1978) and Fig. 5.7. The unpaired electron, which may be present in either the oxidized or reduced form of different Fe/S proteins produces the EPR signal. Each Fe/S group which can be detected by EPR is termed a centre or cluster. A single polypeptide may contain more than one centre. It is not clear how many centres are present in the mitochondrial respiratory chain; complex I may have up to 7 (Section 5.6).

Fe/S proteins contain Fe atoms covalently bound to the apoprotein via a cysteine sulphur and bound to other Fe atoms via acid labile sulphur bridges (Fig. 5.6). Fe/S centres may contain 2 or 4 Fe atoms, even though each centre only acts as a 1 e$^-$ carrier. Fe/S proteins are widely distributed among energy-transducing electron-transfer chains and can have widely different $E_{m, 7}$ values — from as low as -430 mV for chloroplast ferredoxin (Section 6.4) to $+360$ mV for bacterial HiPIP ("high-potential iron–sulphur protein") (Section 6.3).

Ubiquinone (Coenzyme Q, CoQ, UQ or simply Q) is found in mammalian mitochondria as UQ_{10}, i.e. with a side-chain of ten 5-carbon isoprene units (Fig. 5.8). This 50-carbon hydrocarbon side-chain renders UQ_{10} highly hydrophobic. UQ undergoes a $2 H^+ + 2 e^-$ reduction to form UQH_2 (ubiquinol), although there is EPR evidence for a partially reduced free radical form $U\dot{Q}H$ (ubisemiquinone). The role of UQ in the respiratory chain has been a matter of some controversy. The fractional turnover of UQ is low compared to other respiratory chain components, and this led to the quinone being placed on a side-pathway. However, UQ is present in much higher amounts than other respiratory chain components, and the total flux through the UQ pool is adequate for it to be located on the main pathway (Kröger & Klingenberg 1973). The simplest postulate for the role of UQ is as a mobile redox carrier linking complexes I and II with complex III, although in the "Q-cycle" formulation of electron-transfer in complex III, UQ is proposed to play an integral role (Section 5.8). While UQ_{10} is the physiological mediator, its hydrophobic nature makes it difficult to use experimentally, and ubiquinones with shorter side-chains, and consequently greater water solubility, are usually employed.

Some anaerobic respiratory chains employ menaquinone in place of UQ (Section 5.12) while in the chloroplast the corresponding redox carrier (Section 6.4) is plastoquinone (Fig. 5.8).

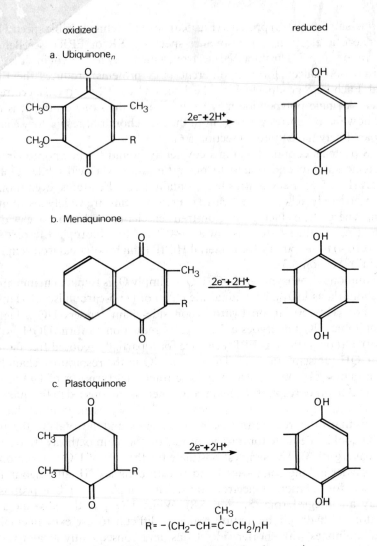

Fig. 5.8   Ubiquinone and related redox carriers.

5.3 THE LINEAR SEQUENCE OF REDOX CARRIERS IN THE RESPIRATORY CHAIN

The sequence of electron carriers in the mitochondrial respiratory chain (Fig. 5.1) was largely established by the early 1960s as a result of the application of oxygen electrode (Section 4.5) and spectroscopic techniques. This work was greatly facilitated by the ability to feed in and extract electrons at a number of locations along the respiratory chain, corresponding to the junctions between the respiratory complexes. Thus NADH reduces complex I, succinate reduces complex II, and tetramethyl-p-phenylenediamine (TMPD) reduces cytochrome oxidase (Fig. 4.2). Ascorbate is usually added together with TMPD to regenerate the reduced form of the dye. Ferricyanide is a non-specific, but impermeant, electron acceptor and can be used not only to dissect out regions of the respiratory chain but also to provide information on the orientation of the components within the membrane (Klingenberg 1979).

The discovery of specific electron-transfer inhibitors enabled the oxygen electrode to be used to localize the relative positions of the sites of electron entry and inhibitor action (Fig. 5.9). Armed with this information, it was possible to proceed to a spectral analysis of the location of each redox carrier relative to these sites.

An independent approach to the ordering of the redox components came with the development of techniques for studying the kinetics of oxidation of the respiratory chain when a transient is induced by the addition of oxygen to an anaerobic suspension of mitochondria (Fig. 5.4). The sequence with which the components become oxidized can reflect their proximity to the terminal oxidase and also whether they are kinetically competent to function in the main pathway of electron transfer. The rapidity of the oxidations observed under these conditions requires the use of stopped-flow techniques (Fig. 5.4).

The carriers in the respiratory chain must be ordered in such a way that their redox potentials, E_h (Section 3.3) form a sequence from NADH to O_2. E_h is determined from the mid-point potential, E_m and the extent of reduction (Eq. 3.19). Although the extent of reduction of a component in the respiratory chain can be measured spectroscopically, indirect methods are needed to measure the mid-point potential *in situ*. It should be noted that the mid-point potential of a component in the respiratory chain is usually different from that of the purified, solubilized component.

The technique of redox potentiometry (see Dutton & Wilson 1976, Dutton 1978) combines dual-wavelength spectroscopy with redox potential determinations (Section 3.3). As with redox potentiometry of most biological couples, it is necessary to add a low concentration of an intermediate redox couple in order to speed the process of equilibration between the platinum electrode and the primary couple. As a secondary mediator will only function effectively in the region of its mid-point potential (so that there are appreciable concentrations of both its oxidized and reduced forms), a set of mediators is required to cover the whole span of the respiratory chain, with mid-point potentials spaced at intervals of about 100 mV. Mediators

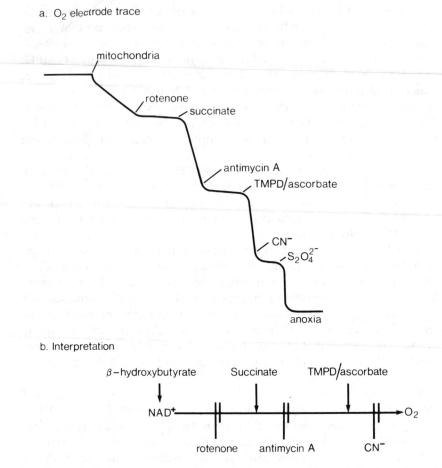

a. O_2 electrode trace

Fig. 5.9 Typical oxygen electrode experiment establishing relative locations of inhibitor sites and points of electron entry into the respiratory chain.

Liver mitochondria were added to an air-saturated medium containing FCCP (a proton translocator to ensure that mitchondrial proton permeability is not rate limiting (Section 2.5.3). The NAD^+-linked substrate β-hydroxybutyrate is initially present. 1 μM rotenone completely inhibits respiration but allows succinate to be subsequently oxidized. Antimycin A inhibits succinate oxidation, but not that due to TMPD plus ascorbate. Cyanide inhibits this respiration. Dithionite ($S_2O_4^{2-}$) is added to induce non-enzymatic anoxia.

are usually employed at concentrations of 10^{-6} to 10^{-4} M. Many mediators are autoxidizable, and the incubation has to be maintained anaerobic for this reason and also to prevent a net flux through the respiratory chain from upsetting the equilibrium.

A second requirement for membrane-bound systems is that the mediators must be able to permeate the membrane in order to equilibrate with all the components. This introduces a considerable complication if the mitochondria are studied under "energized" conditions (i.e. in the presence of ATP since the incubation must be anaerobic). Thus, the $\Delta\Psi$ or ΔpH will affect the distribution of the oxidized and reduced forms of the mediators across the membrane, and the oxidized/reduced ratio of the mediator at the site of the component will differ from that at the platinum electrode. Interpretation of ATP-dependent effects are thus complicated, as the possibility of this artifact must be distinguished from any genuine effect due to redistribution of electrons across the membrane on induction of a membrane potential, or even due to an inherent energy-dependent change in E_m itself (see Hinkle & Mitchell 1970, Walz 1979). The simplest redox potentiometry is therefore performed with de-energized mitochondria or sub-mitochondrial particles.

The practical determination of the E_m of a respiratory chain component is straightforward (Fig. 5.10). Mitochondria are incubated anaerobically in the presence of the secondary mediators. The state of reduction of the relevant component is monitored by split-beam spectroscopy, while the ambient redox potential is monitored by a Pt or Au electrode. The electrode allows the secondary mediators and the respiratory chain components all to equilibrate to the same E_h. This potential can then be made more electronegative (by the addition of ascorbate, NADH, or dithionite) or more electropositive by the addition of ferricyanide, and E_h and the degree of reduction of the component are monitored simultaneously. In this way a redox titration for the component can be established.

Considerable information can be gathered from such a titration. Besides the mid-point potential itself, the slope of log[ox]/[red] establishes whether the component is a 1 e^- carrier (60 mV per decade) or a 2 e^- carrier (30 mV per decade). By repeating the titration at different pH values it can be seen whether the mid-point potential is pH dependent, implying that the component is a $(H^+ + e^-)$ carrier. Finally, the technique frequently allows the resolution of a single spectral peak into two or more components based on differences in E_m. In this case the basic Nernst plot (Fig. 3.3) is distorted, being the sum of two plots with differing E_m values which can then be resolved. One of the most interesting findings with this technique was that cyt b in complex III can be resolved into two components (Section 5.8). Redox potentiometry can also be employed for Fe/S proteins, in which case the redox state of the components is monitored by EPR.

The mid-point potentials of the identifiable components of the respiratory chain are depicted in Fig. 5.11. Once the E_m values have been established for non-respiring mitochondria, an E_h for a component can be assigned to any component in respiring mitochondria simply by determining the degree of

a. Apparatus

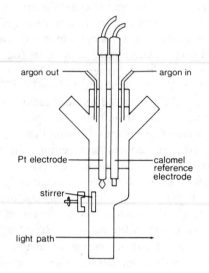

argon out —   — argon in

Pt electrode ——   — calomel reference electrode

stirrer —

light path —

b. Redox difference spectra of succinate–cytochrome c reductase   Complex(II+III)

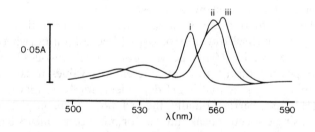

0·05A

ii   iii

i

500   530   560   590

λ (nm)

Fig. 5.10   Redox potentiometry of respiratory chain components.
(*a*) Apparatus for the simultaneous determination of redox potential and absorbance. (*b*) Difference spectra obtained with a suspension of succinate–cytochrome c reductase (i.e. complexes II+III). The complex, held in solution by a low concentration of Triton X-100 and deoxycholate, was added to an anaerobic incubation containing redox mediators. The ambient redox potential was varied by the addition of ferricyanide. (i) Reference scan (baseline) at +280 mV, all cytochromes oxidized, second scan at +145 mV (cyt c_1 now reduced). (ii) Baseline at +145 mV (cyt c_1 reduced), second scan at −10 mV (b_{562} additionally reduced). (iii) Baseline at −10 mV (c_1 and b_{562} reduced), second scan at −100 mV (cyt b_{566} additionally reduced). (Data adapted from Dutton 1978.)

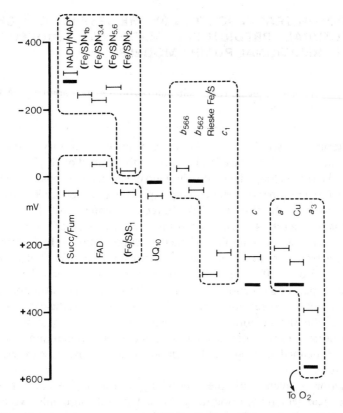

Fig. 5.11 E_m values for components of the mitochondrial respiratory chain and E_h values for State 4.

Values are for heart mitochondria. (————), E_m values obtained with de-energized mitochondria. (▬▬▬) E_h values for mitochondria oxidizing β-hydroxybutyrate in State 4, calculated from the de-energized E_m values and from the extent of reduction of the components in State 4. It was assumed that no energy-dependent change in mid-point potential occurs. (Data adapted from Erecinska *et al.* 1974; Dutton & Wilson 1974.)

reduction. The results for mitochondria respiring in State 4 are shown in Fig. 5.11. The oxido-reduction components fall into four equipotential groups, the gaps between which correspond to the regions where proton translocation occurs. The drop in E_h of the electrons across these gaps is conserved in the proton electrochemical potential.

5.4 PROTON TRANSLOCATION BY THE RESPIRATORY CHAIN; STRUCTURAL PREDICTIONS MADE BY "LOOPED" OR "CONFORMATIONAL PUMP" MODELS

Reviews Greville 1969, Mitchell 1976b, 1979a, Papa 1976, Wikström & Krab 1979

The alternative hypotheses for the mechanism of respiratory chain proton translocation were introduced in Fig. 4.10 in the context of $H^+/2 e^-$ stoicheiometry. The direct looped respiratory chain mechanism proposed by Mitchell is subject to a number of stringent and experimentally verifiable constraints. Proton extrusion by this model is a direct consequence of the transfer of electrons from a $(H^+ + e^-)$ carrier to a pure electron carrier, with the release of the protons on the outer face of the membrane, followed by transfer from a pure electron carrier to a $(H^+ + e^-)$ carrier with the uptake of protons from the matrix (Fig. 4.10). It follows that there is an obligatory stoicheiometry of $2 H^+/2 e^-$ per loop (Section 4.3), that the linear sequence of redox carriers must contain the correct alternation of carriers, and that the carriers should be localized in the membrane in a way which produces the observed orientation of proton translocation. In contrast, a conformational pump is a sufficiently ill-defined concept to be adaptable to almost any experimental observation.

A major problem with the "loop" hypothesis is the lack of evident $(H^+ + e^-)$ carriers in the respiratory chain. Thus there are only two obvious $(H^+ + e^-)$ carriers—FMN in Complex I (Section 5.6) and UQ_{10}, but in contrast there are up to 15 apparent e^--carriers (up to 8 Fe/S proteins, 5 cytochromes and 2 Cu atoms). The protonmotive Q-cycle (Section 5.8) is one solution to this problem, UQ_{10} acting essentially as the $(H^+ + e^-)$ carrier in two loops.

It is sometimes stated that the "loop" hypothesis requires that the loops should span the membrane and that the appropriate redox carriers should be detectable on the outer face of the membrane. However, all that is required for a functional loop is that there should be a means for releasing the proton on the outer face. Thus, all the redox carriers could be located close to the matrix face of the membrane, proton translocation being catalysed by specific proton channels analogous to the F_o component of the ATP synthetase (Section 7.3).

Although it is convenient to categorize theories for respiratory chain proton translocation in terms of loops or conformational pumps, it is important not to exaggerate the differences between the models. Thus the "loop" hypothesis proposes that some redox carriers change their pK upon reduction sufficiently to bind two protons per two electrons from the matrix, and decrease their pK again on reoxidation, releasing their protons to the external phase. This pK change and directionality are also features of any "conformational pump" model, the only distinction being that the constraint that the protons and electrons should bind to the same component is removed. Instead the redox component becomes distinct from the proton-binding component, the Gibbs energy made available by the redox change being transduced, via a conformational change to a second component which responds by altering its pK and conformation. As this second component can be a peptide with no redox properties, the choice of possible candidates is no longer restricted. Secondly, the stoicheiometric restriction is removed, since there is no reason why the pKs of three or more groups should not change in response to the conformational change induced by a $2 e^-$ reduction of the associated redox carrier, whereas no known redox carrier binds more than $2 H^+/2 e^-$.

A conformational pump must co-ordinate a redox change, a conformational change, and a protonation change. Therefore even the simplest model has $2^3 = 8$ hypothetical states interconnected by a cubic array of transitions. Vectorial proton translocation would be imposed by restricting the number of permitted transitions as shown in Fig. 5.12.

5.5 FRACTIONATION AND RECONSTITUTION OF RESPIRATORY CHAIN COMPLEXES

Reviews Hatefi 1978, Hatefi *et al.* 1962 (preparation of complexes); Kagawa 1972, Racker 1979 (isolation and reconstitution)

Bile salts such as cholate and deoxycholate, when used at low temperatures and low concentrations, only disrupt lipid–lipid interactions in membranes, leaving peptide–peptide associations intact. Using these detergents, the mitochondrial respiratory chain can be fractionated into four complexes, termed complex I, II, III and IV (cyt c oxidase). The electron transfer activity of each complex is retained during this solubilization. When the complexes are reconstituted into artificial bilayer membranes, their ability to translocate protons is restored. Fractionation and reconstitution of the complexes serve a number of purposes: (1) the complexity of the intact mitochondrion is reduced; (2) it is possible to establish the minimum number of components which are required for the function of each region of the respiratory chain; (3) during the period in which the chemiosmotic theory was being tested, reconstitution proved one of the most persuasive techniques for eliminating

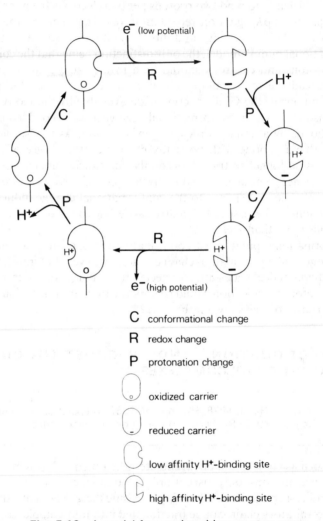

e^- (low potential)

R

H^+

P

H^+

C

R

H^+

H^+

P

C

e^- (high potential)

C   conformational change

R   redox change

P   protonation change

oxidized carrier

reduced carrier

low affinity H^+-binding site

high affinity H^+-binding site

Fig. 5.12 A model for a redox-driven proton pump.
A hypothetical model is shown for a proton pump with a stoicheiometry of
$1H^+/1e^-$. The pump can exist in eight states (the combinations resulting from two
conformations, two redox states and two protonation states). In order to produce a
vectorial proton pumping, only certain transitions between these eight states are
permitted. (Adapted from Wikström & Krab 1979.)

the necessity of a direct chemical or structural link between the respiratory chain and the ATP synthetase. For example, it proved possible to "reconstitute" ATP synthesis by combining such disparate complexes as the beef-heart mitochondrial ATP synthetase and the light-driven H^+-pump from halobacteria (Section 6.6) in a single bilayer.

When the mitochondrial respiratory chain is fractionated, there appear to be two moles of complex IV per mole of complex III. Complex I and complex II are present in substantially smaller molar amounts. The isolated complexes readily reassemble. For example, in the presence of phospholipid and UQ_{10}, complex I and complex III reassemble spontaneously to reconstitute NADH–cyt c oxidoreductase activity (Ragan & Heron 1978).

The techniques for reconstitution have been pioneered by Racker and Kagawa. The technical problems, once the pure complex has been obtained, are considerable. First, a technique must be devised to reintroduce the complex into a bilayer in a way which retains catalytic activity. Secondly, allowance has to be made for the possibility of a random orientation of the reconstituted proton-pumps within the membrane, which would prevent the detection of net transport. Incorporation into vesicles is normally accomplished by suspending the complex in cholate together with phospholipid, and then slowly dialysing away the detergent. Vesicles with the complex embedded in the membrane form spontaneously. Alternatively the complex can be sonicated together with the phospholipid (Section 1.4).

5.6 COMPLEX I (NADH–UQ OXIDOREDUCTASE)

Reviews Ragan 1976 (properties), Hatefi 1978 (preparation)

Complex I catalyses the transfer of 2 electrons from NADH to UQ_{10}, a $\Delta E_{h, 7}$ of some 310 mV (Fig. 5.11). The complex is proton-translocating, as can be shown with intact mitochondria or with reconstituted complex I with oxidized quinone as electron acceptor. Higher H^+/O ratios (Section 4.3) and ADP/O ratios (Section 4.6) are obtained with NAD^+-linked substrates than with succinate, which transfers electrons to UQ_{10} via the non-proton-translocating complex II.

Complex I has a molecular weight of about 850 000, making it probably the largest protein component of the inner membrane. It has at least 16 polypeptides, most of which do not play a direct redox role. NADH is oxidized on the matrix face of the membrane by a FMN-containing component, NADH-dehydrogenase. Intact mitochondria are therefore unable to use added NADH, although inverted sub-mitochondrial particles may. In addition to FMN, complex I contains between 16 and 28 Fe atoms and acid-labile S atoms per mole of FMN, organized in 2 and 4 Fe-clusters. There are between 4 and 7 Fe/S centres per mole of complex I, which have been detected by low temperature EPR in conjunction with redox-potentiometry (Fig. 5.11). With the exception of centre N-2, which has an $E_{m,7}$ of about -80 mV, the centres have negative mid-point potentials varying from -240 to -380 mV.

Complex I may be inhibited by rotenone or piericidin A. The site of rotenone binding is not known, although as FMN and all the Fe/S centres are reduced in the presence of the inhibitor, it is likely that inhibition occurs at the final stage of electron transfer to UQ_{10}.

The non-permeant and non-specific electron acceptor ferricyanide does not interact with complex I in intact mitochondria but catalyses a rapid, rotenone-insensitive NADH oxidation in inverted sub-mitochondrial particles.

5.7 COMPLEX II (SUCCINATE DEHYDROGENASE); ELEC-TRON-TRANSFERRING FLAVOPROTEIN AND α-GLYCERO-PHOSPHATE DEHYDROGENASE

Review Hatefi 1978 (preparation)

In addition to complex I, three other redox pathways feed electrons to UQ_{10} and complex III. These are: complex II, which transfers electrons from succinate; electron-transferring flavoprotein, supplying electrons from the flavoprotein-linked step of fatty-acid β-oxidation; and α-glycerophosphate dehydrogenase, catalysing the extra-mitochondrial oxidation of α-glycerophosphate to dihydroxyacetone phosphate. The first two are located on the matrix face of the membrane, and the other is on the outer face. All three are flavoproteins transferring electrons from substrate couples with mid-point potentials close to 0 mV. As would be expected on thermodynamic grounds, none is proton-translocating.

Complex II consists of up to 4 polypeptides (Table 5.1). The two largest polypeptides constitute succinate dehydrogenase, the largest containing a covalently bound FAD and two Fe/S centres, while the other polypeptide contains an additional centre. Complex II contains cyt b equimolar with FAD, which is distinct from the cyt b of complex III (Section 5.8) and which is associated with the smaller polypeptides. The function of this cytochrome is unclear.

Table 5.1 The functional complexes of the mitochondrial respiratory chain

Complex	Subunits	Prosthetic groups	H^+ translocating?
I NADH–UQ reductase 850 kD	16	1 FMN 16–24 non-haem irons (5–7 centres)	yes
II Succinate–UQ reductase 97 kD	2	1 FAD 8 non-haem irons (3 centres)	no
(Ubiquinone)			
III UQH_2—cyt c reductase 280 kD	6–8	2 b-type haems 1 c-type haem 2 non-haem irons (1 centre)	yes
(Cyt c)			
IV Cyt c oxidase 200 kD	6–7	2 a-type haems 2 Cu	yes[a]

[a] Strictly there is a controversy as to whether electrons are translocated or protons (Section 5.9).
Cytochromes are present in near stoicheiometric amounts, i.e. $b:c_1:c:a:a_3 :: 2:1:2:2:2$.
The complexes are present in the approximate molar ratios $I:II:III:IV::0·1:0·1:0·5:1$

5.8 UBIQUINONE AND COMPLEX III (bc_1-COMPLEX OR UQ-CYT c OXIDOREDUCTASE)

Reviews Hatefi 1978 (preparation); Wikström 1973, Papa 1976, Rieske 1976 (properties)

Although UQ_{10} is the smallest redox component of the respiratory chain, its role has caused perhaps the most controversy. The simplest role for the hydrophobic coenzyme is to act as a "hydrogen-pool", i.e. a mobile ($2 H^+ + 2 e^-$) carrier transferring electrons from complexes I and II to complex III. However, in the "protonmotive-Q cycle" it is proposed that UQ plays an integral role in ($H^+ + e^-$) transfer within complex III itself.

Complex III contains 8 subunits, including 2 moles of cyt b and one mole of cyt c_1. For a long time it was thought that the two b cytochromes were distinct species, as only one b can be reduced by the addition of succinate, the other requiring dithionite. Additionally, the two b cytochromes display dis-

tinct spectra, both in the intact mitochondrion and in the purified complex.
Finally the two bs have different mid-point potentials, a distinction which is
retained even in the purified cyt b dimer (von Jagow et al. 1978). These
apparently distinct species were termed b_K, or b_{562} for the component reducible by succinate alone, and b_T or b_{566} for the dithionite-reducible
component. However, genetic studies with yeast mitochondria have shown
that the apoproteins are coded for by the same structural gene, suggesting the
presence of two structurally identical subunits in the dimer. In this case the
heterogeneous behaviour of the components would be due to a mutual
interaction between the monomers.

The mid-point potentials of b_{566} and b_{562} are about -30 mV and $+30$ mV
respectively (Fig. 5.11). Both components appear to be buried deeply in the
membrane and are not accessible to the non-specific electron acceptor
ferricyanide either in intact mitochondria or inverted submitochondrial
particles. A membrane potential generated either by ATP hydrolysis or by
K^+-diffusion (Section 3.6) causes b_{566} to become reduced relative to b_{562};
the apparent $E_{h, 7}$ of b_{566} goes from -40 mV to $+70$ mV. This suggests that
the two components may be located on opposite sides of the membrane and
that a membrane potential causes the displacement of electrons from b_{562}
on the matrix side to b_{566} on the cytosolic side.

In addition to the b cytochromes, complex III contains cyt c_1, and a single,
very electropositive Fe/S-protein (the Rieske protein), which is a 2Fe/S-
centre. The complex also contains a number of so-called core proteins,
which lack a redox role but may possibly be involved in proton translocation.
A large positive change in E_h occurs between the b cytochromes and cyt c_1,
which indicates that electron flow in this region must be intimately con-
nected with the generation of $\Delta\bar{\mu}_{H+}$.

Cyt c_1 has a molecular weight of 31 000 and has been purified from beef-
heart mitochondria. In contrast to the isolated cytochrome, c_1 in situ is
insensitive to inhibition by trypsin, suggesting that it is an integral protein.
Cyt c_1 has a spectrum which can be distinguished from that of cyt c only in
low temperature spectra (Section 5.2). Cyt c is bound peripherally to the
outer face of the membrane. The fact there is little change in the relative
redox states of cyts c and c_1 when a membrane potential is applied shows
that c_1 is also towards the outer face. The Rieske protein is in rapid equilibrium with cyt c_1.

Complex III is inhibited by antimycin A at very low concentrations, in
fact one mole of inihibtor inhibits one mole of the complex. The site of action
of antimycin A is unclear, indeed the entire pathway of electron flow in
complex III is not agreed upon. The inhibitor enhances the reduction of the
b cytochromes and prevents the reduction of cyt c_1 when electrons are fed
into the complex from UQ. The membrane-potential-induced dismutation

between the reduction of b_{566} and b_{562} is enhanced, while addition of an electron acceptor such as ferricyanide leads not only to the expected oxidation of cyt c_1 but also to an anomalous reduction of the b cytochromes. This curious response is not dependent on $\Delta\Psi$, and can be demonstrated with the solubilized complex (Fig. 5.13). The simplest explanation is that a dismutation of the electron flow occurs under these conditions:

$$[X]_{red} \longleftarrow \begin{array}{c} e^- \nearrow c_1 \\ \\ e^- \searrow b \end{array}$$

If this electron transfer from an unspecified donor is close to equilibrium in the presence of antimycin A, oxidation of c_1 will alter the equilibrium in favour of b reduction.

The lack of a second $(H^+ + e^-)$ carrier and the anomalous behaviour of the cytochromes in the presence of antimycin A prompted Mitchell to modify the classical loop (Fig. 4.10) in favour of the "protonmotive Q-cycle" (Mitchell 1975b) in which UQ plays an integral role, undergoing a two-stage oxidation and reduction via the free radical semi-quinone UQH (Fig. 5.14). The semi-quinone is a highly reactive species which rapidly dismutates into UQ and UQH$_2$; it can, however, be detected by EPR in mitochondria and is conceivably stabilized in a protected environment. The Q-cycle retains the

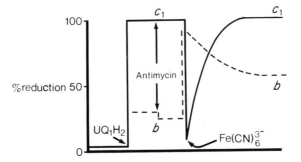

Fig. 5.13 Effects of antimycin A and ferricyanide on the reduction of the cytochromes in purified complex III.
Purified complex III, held in solution by cholate detergent, was reduced by the addition of UQ_1H_2 (the water-soluble analogue of $UQ_{10}H_2$ (Section 5.2)). The quinol completely reduced cyt c_1, but only caused a 30% reduction of b. Addition of antimycin caused little change, but when ferricyanide was added to *oxidize* the complex, c_1, became more oxidized but b became more *reduced*. As the quinol was present in excess over the ferricyanide, these changes reversed as the oxidant was used up. (Adapted from Rieske 1971.)

spatial relationships of the cytochromes arrived at above, while also accounting for the anomalous reduction of b by c_1-oxidation in the presence of antimycin A.

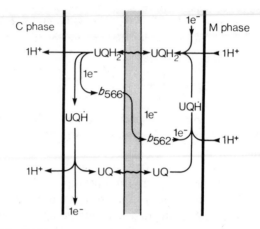

Fig. 5.14   The protonmotive "Q" cycle. (After Mitchell 1975b.)

5.9 CYTOCHROME c AND COMPLEX IV (CYTOCHROME c OXIDASE; FERROCYTOCHROME; O_2 OXIDOREDUCTASE

Reviews Hatefi 1978 (preparation); Wikström & Krab 1979 (proton translocation)

Complex III transfers electrons to cyt c. Cyt c is not isolated as a component of a complex, although it can bind stoicheiometrically to cytochrome c-oxidase. Cyt c is a peripheral protein located on the outer face of the mitochondrial membrane and may be readily solubilized from intact mitochondria. Cytochrome oxidase is the terminal complex of the respiratory chain and catalyses the reduction of O_2 to 2 H_2O in a 4e$^-$ reaction:

$$4\,e^- + 4\,H^+ + O_2 \rightarrow 2\,H_2O$$

The natural electron donor is cyt c. Electrons may also be supplied by the artificial donor tetramethyl-p-phenylenediamine (TMPD) which is in turn reducible by ascorbate (Section 4.1). As befits a 4e$^-$ reaction, four 1-electron redox groups are located in the complex: two haems (cyt a and cyt a_3) and two Cu atoms capable of reduction from Cu^{2+} to Cu^+. The two cytochromes are distinguishable on the basis of the specific ability of a_3 to bind ligands such as CO, although, as with the b cytochromes, it is not certain that they represent chemically distinct species. Unlike the b and c cytochromes, where both of the non-porphyrin co-ordination positions of the Fe are occupied by amino acid side-chains, the a cytochromes resemble haemoglobin in having one free co-ordination position, which enables a_3 to bind O_2 and the inhibitors CO, CN^- and N_3^-.

Electrons enter the complex at an $E_{h,7}$ of about $+290\,mV$ for mitochondria in State 4 (see Fig. 5.11), and are ultimately transferred to the $\frac{1}{2}O_2/H_2O$ couple with an $E_{h,7}$ in air-saturated medium of about $+750\,mV$. The redox span $\Delta E_{h,7}$ is therefore about $460\,mV$. Two electrons falling through this potential would be sufficient to translocate up to 4 protons across the membrane against a $\Delta\tilde{\mu}_{H+}$ of $230\,mV$ (Section 3.8). However, while it is established that complex IV is energy-conserving, there is a lively discussion on the stoicheiometry and mechanism of proton translocation. Unlike the remainder of the respiratory chain, cytochrome oxidase is irreversible.

Two proposals for the mechanism of the complex are depicted in Fig. 5.15. The original proposal of Mitchell is that the complex represents the electron-transferring limb of the final redox loop (Fig. 4.10). In contrast, the proton pump model proposes that a conformational change resulting from electron

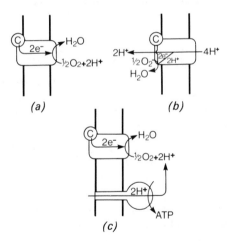

(a)   (b)

(c)

Fig. 5.15 Electron-translocating and proton pumping models for cytochrome oxidase.
(a) Electron-translocating model (Mitchell 1966). Stoicheiometries: $q^+/2e^-=2$; $2H^+/2e^-$ uptake from matrix, no H^+ released to cytosol. (b) Proton pumping model (Wikström & Krab 1979) $q^+/2e^-=4(6)$; $4(6)H^+/2e^-$ uptake from matrix, $2(4)$ $H^+/2e^-$ released to cytosol. (c) Even though no protons are translocated across the membrane in model (a), the electron flow could be coupled to ATP synthesis.

flow through the complex results in the translocation of protons across the membrane. Although no protons are transferred across the membrane in the electron-transferring model, electron flow is sufficient for coupling to ATP synthesis, as protons entering via the ATP synthetase are utilized in the reduction of O_2 to H_2O (Fig. 5.15).

The most apparent difference between the models is that only the proton pump results in the release of protons into the extra-mitochondrial compartment (Fig. 5.15). While it might seem straightforward to establish whether such a release occurs, in practice care must be taken to ensure that any protons observed originate from vectorial proton translocation. Figure 5.16 shows the result of one such experiment, but see Moyle & Mitchell (1978b).

5.10 THE NICOTINAMIDE NUCLEOTIDE TRANSHYDROGENASE

Review Rydström 1977

Although the mid-point potentials for the $NAD^+/NADH$ and $NADP^+/NADPH$ couples are the same, the latter couple is always found to be considerably more reduced in the mitochondrial matrix. This disequilibrium is maintained by an energy-dependent transhydrogenase which catalyses the following reaction:

$$NADH + NADP^+ \rightleftharpoons NADPH + NAD^+$$

The mass–action ratio may exceed 500 and may be maintained either by respiration, in which case oligomycin is without effect, or by ATP hydrolysis, in which case oligomycin is inhibitory. This indicates that the transhydrogenase is dependent on $\Delta\tilde{\mu}_{H+}|$ rather than ATP.

The transhydrogenase has one major polypeptide of 97 000 molecular weight (Höjeberg & Rydström 1979), is membrane-bound, and only responds to matrix nucleotides; it can therefore be studied in inverted sub-mitochondrial particles.

Evidence that the transhydrogenase involves the movement of charge across the membrane was obtained by following the movements of lipophilic anions (Section 2.5) across the membrane of sub-mitochondrial particles (Fig. 5.17). The transhydrogenase is an interesting exception to the rule that there should be a difference in mid-point potential across an energy-transducing step (Fig. 5.11).

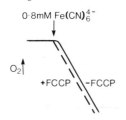

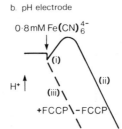

c. Initial ion movements

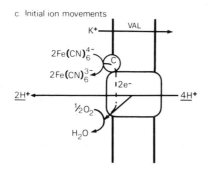

d. Steady state ion movements

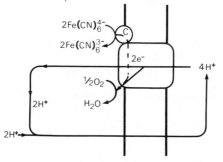

Fig. 5.16 Proton translocation by cytochrome oxidase.
Rat liver mitochondria were added to a medium containing lightly buffered KCl
(as osmotic support), valinomycin (for charge neutralization during initial H^+-
extrusion) and rotenone plus antimycin A (to ensure that no other electron transfer
can occur in other regions of the respiratory chain). Ferrocyanide was added to
reduce cyt c, and following an addition artifact an acidification of the medium was
seen (i) which turned into a steady alkalinization as protons leaked back into the
matrix. In the presence of a proton translocator (iii), only H^+ uptake was seen. (a)
O_2 electrode trace, (b) pH electrode trace, (c) initial conditions, (d) steady state
conditions. (From Wikström & Krab 1978.)

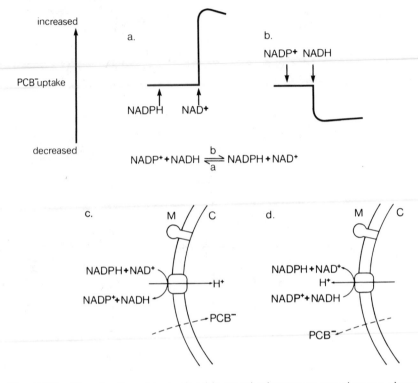

Fig. 5.17 The nicotinamide nucleotide transhydrogenase translocates charge across the membrane.

Submitochondrial particles were incubated in buffered sucrose containing rotenone and CN^- to inhibit the respiratory chain. 10^{-8} M PCB^- (phenyldicarbaun-decaborane, a lipophilic anion) was present, and changes in external PCB^- concentration were followed with a black lipid membrane. In (*a*) the transhydrogenase runs in reverse and pumps protons from the M-phase to the C-phase. In (*b*) the transhydrogenase translocates protons into the M-phase while reducing $NADP^+$ to NADPH. In (*c*) and (*d*) the respective ion movements are shown. (From Skulachev 1970.)

5.11 THE RESPIRATORY CHAIN OF PLANT MITOCHONDRIA

Reviews Palmer 1979, Moore & Rich 1980

Plant mitochondrial respiration is only partially inhibited by CN^-, suggesting the presence of an additional terminal oxidase. In the case of mito-

chondria isolated from the mature spadix of *Arum maculatum* the
CN^--resistant pathway is an order of magnitude more active than the conventional cyt c oxidase, Fig. 5.18. The nature of this terminal oxidase is still

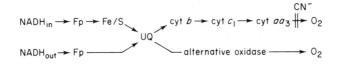

Fig. 5.18 Pathways of electron transfer in plant mitochondria.

unclear, although at various times suggestions have been made of flavoproteins, autoxidizable b-type cytochromes or Fe/S proteins. In the *Arum*
mitochondrion, the pathway to the alternative oxidase appears to diverge at
the level of UQ; for example it is insensitive to antimycin A. The alternative
pathway is not proton-translocating, so that the efficiency of energy-
transduction is low. The consequent heat production is of physiological
importance for distilling insect-attractants to aid pollination. This thermogenic mechanism makes an interesting contrast with that evolved by
mammalian brown fat mitochondria (Section 4.5), in which proton translocation is normal, but a dissipative proton re-entry exists.

Plant mitochondria are also distinctive in their ability to oxidize exogenous
NADH, while at the same time the oxidation of endogenous NADH is only
partially inhibited by rotenone. The two pools of NADH are oxidized by
distinct enzymes; oxidation of exogenous NADH is activated by Ca^{2+} and
inhibited by external AMP or the oxidation of succinate. The external pathway is not proton translocating.

The presence of two pathways to UQ, and two pathways from UQ to O_2
raises the possibility that there are two complete pathways operative, which
are independent but which communicate via the quinone pool (Fig. 5.18).

5.12 BACTERIAL RESPIRATORY CHAINS

Reviews Jones, C. W. 1977, Haddock & Jones 1977, Kröger 1978,
Haddock 1980

Compared with mitochondria, bacterial respiratory chains present three
additional degrees of complexity. First, bacteria possess a variety of electron
transfer chains. Secondly, a given bacterium may change the nature of its

respiratory chain in response to changes in growth conditions. Thirdly, many bacterial respiratory chains are branched, with alternative electron transfer pathways to multiple terminal electron acceptors. This complexity reflects the great versatility of bacteria. For example *E. coli* possesses a linear respiratory chain when grown under highly aerobic conditions, with cyt *o* as terminal oxidase. Under conditions of oxygen limitation a second oxidase, cyt *d*, is synthesized, and electron flow is divided. Under strictly anaerobic conditions electrons can be transferred to nitrate or fumarate. A further degree of versatility is shown by non-sulphur purple bacteria which possess a cyclic photosynthetic electron transfer pathway in the light and a respiratory chain to O_2 in the dark, the two sharing a number of common components (Section 6.3).

The respiratory chains of aerobically grown bacteria employ redox carriers similar but not identical to those used by mitochondria, including Fe/S-proteins, flavoproteins, quinones and cytochromes. However, in only a few cases, such as *Paracoccus denitrificans* (John & Whatley 1977) do the order and complement of the electron carriers become very similar to that of the mitochondrion. The close homology in this case raises the possibility that mitochondria might have originated by endosymbiosis of an ancestor of *Paracoccus*.

In Gram-positive bacteria UQ is replaced by menaquinone (Fig. 5.8), while in many aerobic respiratory chains cyt aa_3 is replaced as terminal oxidase by cyt *o*, or the more cyanide-resistant cyt *d*. Another common variation from the mitochondrial respiratory chain is the absence of a high-potential *c*-type cytochrome. Bacteria such as *E. coli* which are incapable of synthesizing cyt *c* translocate fewer protons in the span from quinone to O_2 (Fig. 5.19).

Proton-translocating regions of the *E. coli* respiratory chain can be divided into low- or high-potential segments. Low-potential segments, Fig. 5.19, transfer electrons from low-potential couples such as $NAD^+/NADH$, $2H^+/H_2$ or $HCOO^-/CO_2$ to menaquinone ($E'_m = -74$ mV) under anaerobic conditions, or to ubiquinone (UQ_8 rather than the mitochondrial UQ_{10}) under both aerobic and anaerobic conditions.

The NADH-dehydrogenase contains FMN and Fe/S centres, as in the mitochondrion. Hydrogenase is an Fe/S protein; there is disagreement on the location of the catalytic site, if on the outer face proton translocation becomes an example of a "Lundegårdh mechanism" (Fig. 1.14). Formate dehydrogenase is a molybdenum-, Fe/S- and haem-containing complex. All these dehydrogenases are proton translocating. The menaquinone reduced under anaerobic conditions can be re-oxidized by fumarate.

Although the reaction catalysed by fumarate reductase is merely the reverse of succinate dehydrogenase, the forward and reverse reactions in *E.*

a. Low potential

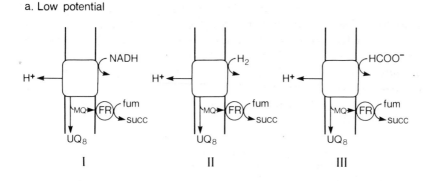

I   II   III

b. High potential

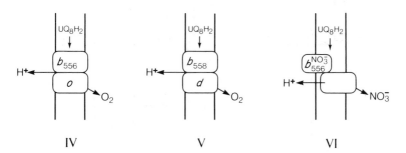

IV   V   VI

Fig. 5.19 Proton-translocating segments of the *E. coli* respiratory chain. (I) NADH-dehydrogenase; (II) hydrogenase; (III) formate dehydrogenase; (IV) cyt *o*; (V) cyt *d*; (VI) nitrate reductase. FR, fumarate reductase. Proton stoicheiometry is not specified. (After Haddock 1980.)

coli are catalysed by separate enzymes. However, both contain FAD as prosthetic group.

The electron donor to the high potential segments of the *E. coli* respiratory chain is UQH_2, which in turn is reduced either by the low-potential segments described above, or by donors such as D-lactate or succinate. Under conditions of high O_2 tension, electrons are transferred via cyt b_{556} to cyt *o*, which is an autoxidizable *b*-type cytochrome. When O_2 is limiting, a cyt b_{558} transfers electrons to cyt *d* (which contains a partially saturated chlorin ring in place of porphyrin). Under strictly anaerobic conditions and in the presence of NO_3^-, nitrate reductase is synthesized, which is a molybdenum-containing, Fe/S protein associated with a specific cyt $b_{556}^{NO_3^-}$ (Fig. 5.19). All these segments are proton translocating.

Chemolithotrophic bacteria use inorganic electron donors and can grow in a completely inorganic mineral salt medium supplemented with CO_2 and a nitrogen source. The terminal electron acceptor is usually O_2, or in a few cases NO_3^-. The $E_{m,\,7}$ values of many of the electron-donating couples is very high, and reduction of NAD^+ in these bacteria presumably occurs by reversed electron transfer (Section 4.7). An extreme case is *Thiobacillus ferrooxidans*, which can use the Fe^{3+}/Fe^{2+} couple at pH 2 as electron donor ($E_m = +780$ mV). Electron transfer to O_2 has been proposed to be another example of a "Lundegårdh mechanism" (Section 1.4; Fig. 5.20).

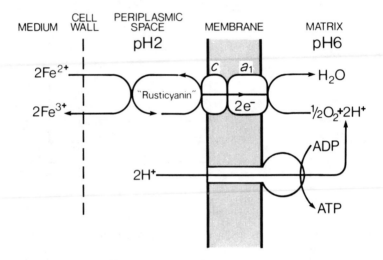

Fig. 5.20 Oxidative phosphorylation in a chemolithotroph, *Thiobacillus ferrooxidans,* which grows at pH 2 using Fe^{2+} oxidation by H_2O as the sole energy source for growth and CO_2 fixation. The Fe^{2+}/Fe^{3+} couple has an E_m of $+780$ mV. (Data from Ingledew *et al.* 1977.)

Spinach chloroplasts are subjected to an acid bath (Fig. 4.19) in order to generate ATP in the dark, while the "Z" scheme of non-cyclic electron transfer generates O_2 and transfers electrons to a negative E_o

6 Photosynthetic Generators of Proton Electrochemical Potential

6.1 INTRODUCTION

The generation of ATP by photosynthetic energy-transducing membranes involves a proton circuit which is closely analogous to that already described for mitochondria. In both cases a $\Delta\tilde{\mu}_{H^+}$ in the region of 200 mV is generated across a proton-impermeable membrane, and this is used to drive a proton-translocating ATPase in the direction of ATP synthesis. The ATPase (or ATP synthetase) is identical to the mitochondrial enzyme except in detail (Section 7.2). The distinction between the two systems comes of course in the nature of the primary generator of $\Delta\tilde{\mu}_{H^+}$, yet even here a number of familiar components recur, including cytochromes, quinones and Fe/S centres.

The two features that are unique to photosynthetic systems are the antennae, responsible for trapping photons, and the reaction centres (RC), to which the light energy is directed. The RCs accept electrons at a positive potential, become excited by the absorption of a photon, and release an electron at a potential which is up to 1 volt more negative. In this way light energy is directly transduced into redox potential energy (Fig. 6.1). In the case of photosynthetic bacteria, this "energetic" electron is fed into an electron transfer pathway which returns the electron to the RC, at the same time using the redox potential to pump protons. This cyclic electron transfer is modified in chloroplasts to become a non-cyclic pathway in which electrons are extracted from water, pass through a RC, an electron transfer chain and a second RC and are ultimately donated to $NADP^+$, at a redox potential 1.1 volt more negative than the $\frac{1}{2}O_2/H_2O$ couple (Fig. 6.1). The chloroplast not only accomplishes this "uphill" electron transfer but at the same time generates the $\Delta\tilde{\mu}_{H^+}$ for ATP synthesis. The ATP and NADPH are used in the Calvin cycle, the dark reactions of photosynthesis in which CO_2 is fixed.

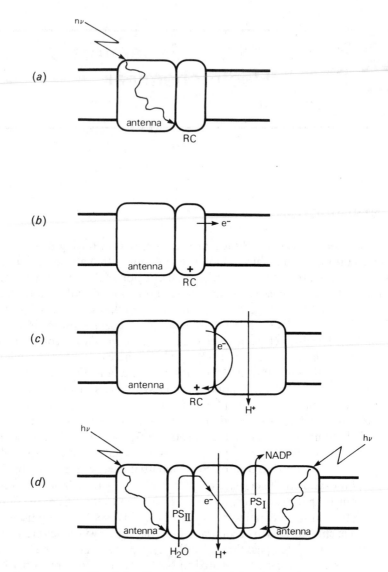

Fig. 6.1 Photosynthetic electron transfer.
In (a) the antenna absorbs photons and transfers the excitation energy to the reaction centre (RC). In (b) the RC ejects an electron at a negative potential. In bacteria (c) the electron cycles back to the RC through a proton-translocating complex similar to mitochondrial complex III (Section 5.8). In chloroplasts (d) electron transfer is non-cyclic. Electrons are extracted from water by one reaction centre (PS_{II}), transferred via a proton-translocating electron-transfer chain to a second reaction centre (PS_I) and ultimately to $NADP^+$.

6.2 THE LIGHT REACTION OF PURPLE BACTERIA

Reviews Parson & Cogdell 1975, Blankenship & Parson 1978, Prince & Dutton 1978

The heavily pigmented membranes of photosynthetic organisms act as antennae, absorbing light and funnelling the resultant energy to specific reaction centres (RC). The interaction of the photon with the reaction centre causes a component to undergo a large change in mid-point potential to a more negative value. As a result an electron can be donated to an acceptor which is up to 1 volt more negative than the potential at which the electron entered the RC. The equivalent energy of a 870 nm photon amounts to 1·42 eV (Section 3.7); thus the energy transfer process is highly efficient.

For every RC, there may be between 30 and 3000 antennae chlorophyll, and the problem of detecting spectral changes due to the RC against the background absorption of the antennae pigments would be considerable were it not for the possibility of purifying RCs free of antennae while retaining the primary light reaction. RCs have been purified from the chromatophores of a variety of purple bacteria (see Gingras 1978). The method involves detergent solubilization of the membrane, followed by conventional protein purification techniques.

While RCs of different purple bacteria vary in detail, typically they have molecular weights of about 75 000 and consist of 2 or 3 polypeptides with 4 molecules of bacteriochlorophyll (Bchl), 2 molecules of bacteriopheophytin (Bpheo), 1 Fe atom, and 1 or 2 molecules of ubiquinone.

The primary photochemical event in the RC is the change in E_m of a dimer of Bchl molecules (referred to as P, often followed by the wavelength of the major absorption band, e.g. P_{870}). P in the non-excited state has an $E_{m,7}$ of $+470$ mV, and the change in E_m occurs when the dimer becomes excited by the absorption of a quantum. Excitation of P to P* occurs in less than 10^{-15}s and results in the shift of electrons to higher energy orbitals. This in turn increases the ease with which an electron can be lost from P*, i.e. it causes a negative shift in E_m. Following the loss of an electron from P*, the resulting unpaired electron of P^+ is delocalized over the two molecules of Bchl (Fig. 6.2).

The immediate acceptor of the electron is a Bpheo molecule "I"). Bpheo is a chlorophyll derivative, where the Mg^{2+} is replaced by two protons. The transfer of an electron from P* to I can be detected in less than 10 ps, and the resulting $(Bchl)_2^+ \ldots (Bpheo)^-$ biradical (termed P^F) has a characteristic spectrum. Studies of isolated Bpheo in hydrophobic solvents suggest that its

a. Photochemical events

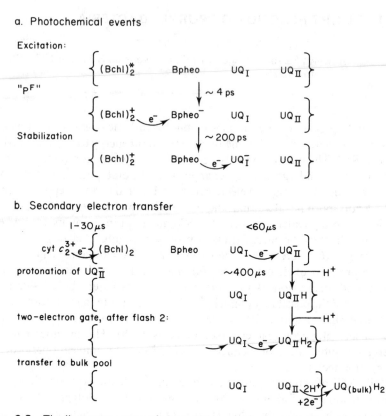

Fig. 6.2 The linear sequence of electron transfer in the bacterial reaction centre.

E_m in the RC is about -550 mV, or more than 1 volt more negative than P in its unexcited state (Fig. 6.3).

The biradical P^F is highly unstable, and within 200 ps, the electron is further transferred to an acceptor "X", which in the case of *Rhodopseudomonas sphaeroides* is one of the two bound UQ molecules ("UQ_I"). The addition of a single electron to UQ_I results in the formation of an anionic semiquinone, $UQ_{\overline{I}}$. The effective E_m of the $UQ_I/UQ_{\overline{I}}$ couple is about -180 mV. The electron is further transferred to the second bound quinone, UQ_{II} before UQ_I has had time to become protonated to form UQ_IH.

UQ_I thus oscillates between the oxidized and anionic semiquinone forms and never becomes fully reduced. UQ_{II} in contrast becomes protonated by 400 μs after accepting an electron. Furthermore, UQ_{II} appears to be unable to transfer a single electron to the bulk pool of UQ located outside the RC. Instead it waits until a second electron, generated by another quantum,

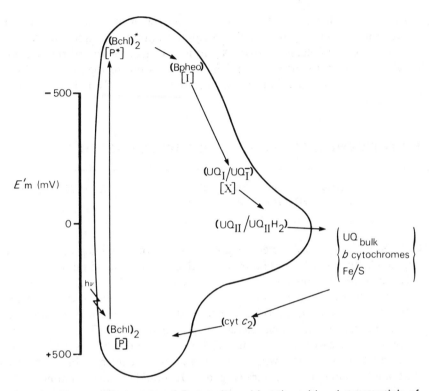

Fig. 6.3 The reaction centre of *Rps. sphaeroides*, the mid-point potentials of the intermediates. (After Dutton & Prince 1978.)
It should be noted that equilibrium values obtained by redox potentiometry (Section 5.2) are not necessarily applicable during the very rapid electron transfer in the RC, for example the UQ_I/UQ_I^- couple is only stable in the nsec time-scale of the light reaction.

reduces $UQ_{II}H$ to $UQ_{II}H_2$ (a second protonation occurring). The fully reduced bound quinone can then transfer $2\,e^- + 2\,H^+$ to the bulk pool of UQ. The two bound UQ molecules thus act as a two-electron gate, transducing the one-electron photochemical event into a $2\,e^-$ transfer. As will be seen in Section 6.3, the protonations of the bound UQ_{II} play an essential role in the generation of $\Delta\tilde{\mu}_{H+}$.

While these events are occurring at the acceptor end of the RC, electrons must be fed into the RC in readiness for the next photochemical event. This is accomplished by cyt c_2, which donates an electron to P^+ to regenerate P. The cyclic electron transfer is then completed by a pathway from the bulk UQ back to cyt c_2; this will be discussed in Section 6.3.

An essential feature of the primary photochemical event is its virtual irreversibility. Due to the fall in potential from P* through I to X, the reverse transfer of electrons from X^- to P^+ occurs at least 10^4 times more slowly than the forward reaction. This is the reason for the almost perfect quantum yield, i.e. one photon results in the creation of one low-potential electron.

6.3 THE GENERATION OF $\Delta\bar{\mu}_{H^+}$ IN PURPLE BACTERIA

Reviews Gromet–Elhanan 1977, Jones, O. T. G. 1977, Dutton & Prince 1978, Wraight *et al.* 1978

Carotenoids are a very heterogeneous class of long-chain, predominantly aliphatic pigments which are found in both chloroplasts and photosynthetic bacteria. Over 350 distinct compounds have been identified. A common feature of these molecules is a central hydrophobic region of the molecule with conjugated double bonds allowing delocalization of electrons and giving the carotenoids a characteristic visible spectrum. The most useful property of the carotenoids is their ability to respond to the extremely high electrical field generated within an energy-transducing membrane with a shift of a few nm in their spectra (Fig. 6.4). It should be borne in mind that a membrane potential of only 100 mV across a membrane 10 nm thick implies an average field strength of 10^5V cm^{-1}. This band shift is not restricted to carotenoids but can also be detected to a lesser extent in chlorophyll. Nor is

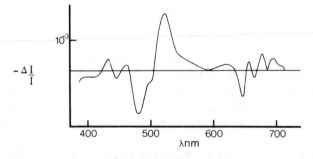

Fig. 6.4 Light-induced electrochromic absorption changes in chloroplasts. (After Junge 1977.)
The difference spectrum (Section 5.2) created by the light-induced red shift in the absorption spectra of the pigments is plotted. The large peak at 518 nm is mainly due to carotenoids, hence the "carotenoid band shift".

it only apparent in bacteria; indeed "electrochromic" effects were first described in chloroplasts (Junge & Witt 1968).

While carotenoids are not essential to photosynthetic processes, the extreme rapidity of their response to illumination of photosynthetic membranes (less than 20 ns) suggests that they are registering some primary charge-transfer event in the membrane. Indeed the initial phase of the carotenoid response can be correlated with the transfer of electrons from P* to X. This implies that the primary photochemical reaction is electrogenic, i.e. associated with charge transfer across the membrane, although it should be remembered that carotenoids detect a local field within the membrane, which need not be the same as a bulk phase potential difference across the membrane (Section 3.6). A carotenoid band shift in the same direction as that produced in the light can be generated in the dark by a K^+-diffusion potential in the presence of valinomycin, the interior space of the chromatophore being positive.

Much evidence suggests that the RC plugs through the membrane (Fig. 6.5). The immediate donor to P, cyt c_2, is located on the inner face of the chromatophore, corresponding to the periplasmic face of the membrane in the intact bacterium (Section 1.4), and can be released from intact bacteria by digestion of the cell wall. In contrast, the protons taken up by UQ_{II} when it becomes reduced (Section 6.2) are taken from the external medium in chromatophore preparations (Fig. 6.5).

The carotenoid band shift can detect two further electrogenic steps in addition to the transfer of electrons from P to X. One correlates with the transfer of an electron from cyt c_2 to P; the other is related to an event in the pathway from UQ_I back to cyt c_2 (see below).

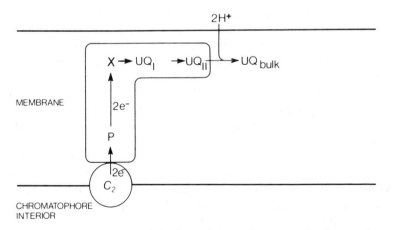

Fig. 6.5 The organization of the bacterial reaction centre in the membrane.

After a single flash, the carotenoid band shift disappears with a time course which correlates with the re-equilibration of protons across the membrane and can be greatly accelerated by the addition of a proton translocator. Thus even though carotenoids detect local fields in the membrane, they respond to the decay of the bulk phase potential gradient.

The pathway of electron transfer from the bulk pool of UQ back to cyt c_2 is controversial. In the case of *Rps. sphaeroides*, the redox carriers available include 3 Fe/S proteins, and 2 or 3 b-type cytochromes which can be distinguished on the basis of their $E_{m, 7}$ (respectively $-90\,mV$, $+50\,mV$ and $+15\,mV$. Any pathway has to fit with two observations. First, in the presence of valinomycin (to slow down the build-up of $\Delta\Psi$) a $2H^+/e^-$ stoicheiometry of proton uptake by chromatophores can be seen during repetitive single turnover flashes. Secondly, analysis of the carotenoid band shift in response to a single turnover flash reveals a third, slow electrogenic component. The antimycin A sensitivity of this component suggests, by analogy to the mitochondrion (Section 5.8), that b-type cytochromes are involved. These observations do not fit with the simplest way of completing cyclic electron flow, diffusion of UQH_2 across the membrane and direct reduction of cyt c_2 with the discharge of protons into the internal space of the chromatophore, because this hypothetical pathway would involve no slow electrogenic step and would give a H^+/e^- ratio of 1. It would also be inefficient: the redox drop between UQ and cyt c_2 is adequate for a second "energy-transducing" (i.e. proton-translocating) step.

The redox carriers which are available are strikingly similar to those present in the mitochondrial proton-translocating complex III (Section 5.8). As with the mitochondrial system the problem arises that there is a lack of sufficient obvious $(H^+ + e^-)$ carriers to translocate the protons, and a "Q-cycle" (Section 5.8) has been proposed, involving a hypothetical $(H^+ + e^-)$ carrier "Z", and a two-stage reduction of the bulk UQ pool via the semiquinone (Fig. 6.6). It should be emphasized that neither of the models shown in Fig. 6.6 is adequate to explain all the data and the actual pathway is still not established.

Cyclic electron transfer generates $\Delta\tilde{\mu}_{H+}$ but does not produce reducing equivalents for biosynthesis. In purple sulphur bacteria, these electrons are obtained from electron donors such as H_2S or $S_2O_3^{2-}$, while for non-sulphur purple bacteria the electron donors include malate and succinate. As the redox potentials achieved by the components of the cyclic electron transfer pathway are all more positive than that of the $NAD(P)^+/NAD(P)H$ couple, nicotinamide nucleotides cannot be reduced by direct electron flow as occurs in chloroplasts (Section 6.4). Instead a $\Delta\tilde{\mu}_{H+}$-dependent reversed electron transfer (Section 4.7) is involved (Fig. 6.7).

Most members of Rhodospirillacae can grow aerobically in the dark.

a. 'Second–loop' scheme involving a $(H^+ + e^-)$–carrier 'Z'
(see Wraight et al. 1978)

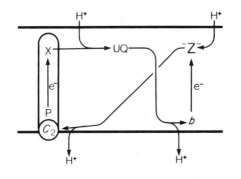

b. 'Q–cycle' scheme (see Dutton & Prince 1978)

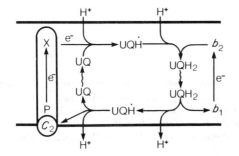

Fig. 6.6 Possible proton-translocating electron-transfer pathways in purple bacteria.

Oxygen represses the synthesis of bacteriochlorophyll and carotenoids, and so the RC is absent. However, the b and c cytochromes are retained, and a terminal oxidase is induced, which in the case of *Rps. sphaeroides* is a Cu-containing protein very similar to the mitochondrial complex IV (Section 5.9). By using the pre-existing cytochromes, the induced oxidase and the reversed electron transfer pathway from quinone to NAD(P)H, the bacterium can therefore assemble a complete respiratory chain (Fig. 6.7). The similarity between the redox carriers of photosynthetic electron transfer and mitochondrial complex III is therefore explained. Cyclic photosynthetic electron transfer may be reconstituted in aerobically grown cells by the addition *in vitro* of reaction centres and antennae complexes.

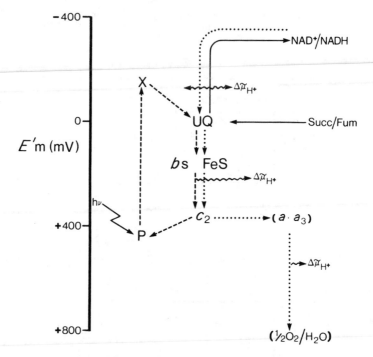

Fig. 6.7   Pathways of electron transfer in *Rps. sphaeroides*.
(— — — →) electron flow during cyclic electron transfer. (————→) $\Delta\tilde{\mu}_{H^+}$
dependent transfer of electrons from succinate to NAD^+. (- - - - - - →) aerobic
respiratory chain following induction of terminal oxidase. A simplified scheme is
shown, with no attempt to resolve electron flow in the region of the *b* cytochromes.

6.4 THE ELECTRON-TRANSFER PATHWAY IN CHLOROPLASTS

Reviews Hill 1965, Trebst 1976, Crofts & Wood 1977, Hauska &
Trebst 1977, Junge 1977, Blankenship & Parsons 1978, Witt 1979

Photosynthetic electron transfer in chloroplasts differs from that in purple
bacteria in two respects: (1) it is non-cyclic, resulting in a stoicheiometric
oxidation of H_2O and reduction of $NADP^+$; (2) two independent light reactions act in series to encompass the redox span from $H_2O/\frac{1}{2}O_2$ to $NADP^+/$
NADPH (Fig. 6.1).

The presence of two RCs within a single membrane has hindered their
purification, and this in turn has meant that information has had to be

gained from intact thylakoid membranes where less than 1% of the pigments are involved in the photoreaction, rather than acting as antennae (Bennett 1979). Highly sensitive techniques are therefore required.

The electron transfer pathway is summarized in Fig. 6.8. The primary electron donor of photosystem I is called P_{700} and appears to be a dimer of chl-a with many similarities to the bacterial P_{870}. The mid-point potential of the unexcited P_{700} is about $+450$ mV. Upon excitation the mid-point potential becomes more negative (Fig. 6.9) and an electron is ejected. The carotenoid band shift (Section 6.3), which was first detected in chloroplasts, indicates that the electron is transferred across the membrane in less than 20 ns. The

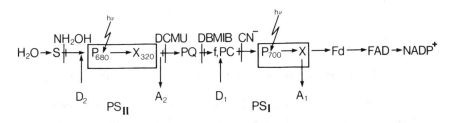

Fig. 6.8 The linear sequence of electron transfer in chloroplasts. Electron-transfer inhibitors: NH_2OH, DCMU (N-dichlorophenyl-N'-dimethylurea), DBMIB (dibromomethyl-isopropyl-p-benzoquinone) and CN^- act where shown. D_2: donors to photosystem II (e.g. benzidine, catechol). A_2: acceptors from photosystem II (e.g. ferricyanide, silicomolybdate, phenylenedi-amines, benzoquinones, DAD [diaminodurane]). D_1: donors to photosystem I, e.g. reduced phenylene diamines, $DPIPH_2$ (reduced 2,6-dichlorophenolindophenol). A_1: acceptors from photosystem I, e.g. ferricyanide, $NADP^+$, MV (methyl-viologen), AQ (anthraquinone-2-sulphonate). (Data from Hauska and Trebst 1977).

nature of the primary acceptor is unclear, but bound ferredoxin, with an $E_{m, 7}$ of -530 mV, is rapidly reduced. Electron transfer to $NADP^+$ is completed by a ferredoxin-$NADP^+$ reductase.

The re-reduction, within 200 μs, of the primary donor, $P700^+$ is accomplished by plastocyanine (PC), a Cu-protein which is a 2 e^- acceptor with an E_m of $+370$ mV. The role of cyt f in this process is controversial.

Photosystem II possess the remarkable ability to extract electrons from H_2O ($H_2O/\frac{1}{2}O_2$ $E_{m, 7} = +820$ mV). The primary electron donor of PS_{II} is designated P_{680} and is probably a chl-a. As with PS_I, an electric field is generated in less than 20 ns. The primary elelctron acceptor is a plasto-quinone X_{320} which resembles UQ_I of the bacterial system in alternating between the fully oxidized and semiquinone anion states (PQ^-). As with the

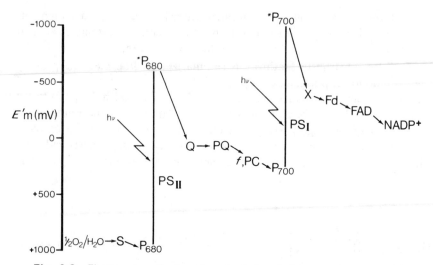

Fig. 6.9 Electron transfer in chloroplasts in relation to redox potential.

bacterial RC, a second bound PQ may act as a 2 e^- transducer before releasing electrons to the bulk pool. No intermediate between P_{680} and X_{320} has yet been detected.

The ultimate electron donor to PS_{II} is H_2O. An enzyme system designated "S" has been proposed to lose 4 electrons to P_{680} prior to oxidizing 2 H_2O to $O_2 + 2 H^+ + 4 e^-$.

The link between PS_{II} and the plastocyanine which reduces PS_I is formed by a bulk pool of PQ which also serves both as an "electron buffer" between the photosystems and also, probably, as a Mitchellian H-limb for the transport of 2 $H^+/2 e^-$ into the thylakoid space (Fig. 6.10).

As with the mitochondrial complex III (Section 5.8) and electron transfer in purple bacteria (Section 6.3), less straightforward roles for the quinone have been proposed (see Hauska and Trebst, 1977), including the involvement of UQ together with the b-type cytochromes in cycles analogous to the mitochondrial Q-cycle (Section 5.8).

The carotenoid response indicates that both the photosystems are oriented across the membrane. The ability to accumulate lipophilic anions (Section 2.5.4) within the thylakoid space during illumination means that the photosystems transport electrons outwards and generate an electrical potential positive in the thylakoid space. Further evidence for the orientation of PS_{II} comes from the observation that the proton liberated in the cleavage of H_2O is only detected in the medium when a proton translocator is added, indicating that water reduction occurs on the inner face. Also, X_{320} must be located close to the outer face, since it can be made accessible to impermeant electron

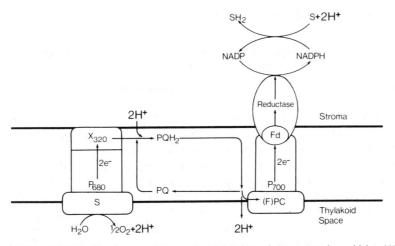

Fig. 6.10 Location in the membrane of chloroplast electron carriers. (After Witt 1979.)

acceptors such as ferricyanide after brief trypsin treatment. Ferredoxin and ferredoxin-NADP$^+$-reductase are accessible to added antibodies, which suggests that PS_1 is similarly oriented (Fig. 6.10).

While the natural electron transfer pathway in chloroplasts is non-cyclic, artificial cyclic electron transfer may be induced by agents such as phenazine methosulphate (PMS), which are capable of acting as both donors and acceptors for PS_I, being reduced on the outer face by PS_I, diffusing through the membrane and then donating electrons back to PS_I on the internal face. If the agent is a $(H^+ + e^-)$ carrier, this creates an artificial loop across the membrane which can lead to cyclic photophosphorylation (Fig. 6.11).

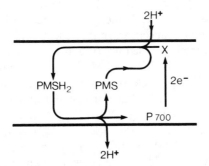

Fig. 6.11 Artificial proton-translocating cyclic electron transfer in chloroplasts. PMS, PMSH$_2$: oxidized and reduced forms of N-methyl-phenazonium methosulphate.

6.5 THE PROTON CIRCUIT IN CHLOROPLASTS

Reviews Jagendorf 1975, Trebst 1976, Avron 1977, 1978, Hauska &
Trebst 1977, Reeves & Hall 1978

The orientation of the electron-transfer pathway and the ATP synthetase
(Section 1.3) indicates that the chloroplast proton circuit operates in the
opposite sense to that of intact mitochondria. Chloroplasts with broken envelopes therefore take up protons when illuminated. The initial $\Delta\Psi$ resulting
from the light reactions is converted into a ΔpH over a time-course of
seconds as a consequence of the redistribution of Mg^{2+} (in intact chloroplasts) or Cl^- (in broken systems) across the thylakoid membrane (Fig. 4.8).
In the steady-state, ΔpH can exceed 3 pH units, as estimated from the accumulation of radio-labelled amines or from the quenching of 9-amino-
acridine fluorescence. The transient $\Delta\Psi$ decays too rapidly to be measured by
radio-labelled anion distribution, but can be followed from the decay of the
carotenoid shift.

The initial proton movements upon illumination can be followed spectro-
photometrically by pH-sensitive dyes such as Neutral Red. Although the
transfer of an electron from PS_{II} to PQ is complete within 2 ms, the time-
constant for the disappearance of protons from the medium is 60 ms,
suggesting that there is an appreciable diffusion-limited barrier for protons
on the outer face of the thylakoid membrane.

As with mitochondria there are considerable problems with the determination of $H^+/2e^-$ ratios in chloroplasts, and there is a consequent lack of con-
sensus in the literature. The "Z" scheme (Figs 6.1 & 6.10) predicts that for
every 2 electrons transferred, 2 protons appear in the thylakoid space due to
the photolysis of water, and 2 protons disappear from the medium due to the
formation of PQH_2. The stoicheiometry associated with PS_I will depend on
the nature of the terminal acceptor. In the case of a $(2 H^+ + 2 e^-)$ acceptor,
2 protons disappear from the medium (Fig. 6.10); stoicheiometric reduction
of the natural acceptor, $NADP^+$ (a$(1 H^+ + 2 e^-)$ acceptor) leads to the uptake of 1 proton, while a pure electron acceptor such as ferricyanide leads to
no proton uptake. In each case 2 protons appear in the thylakoid space due
to the re-oxidation of PQH_2 (Fig. 6.10).

The chloroplast ATP synthetase (Section 7.2) is similar to the mitochondrial enzyme, and ATP-dependent H^+-uptake can be observed in the
dark following activation of the latent enzyme by an imposed $\Delta\tilde{\mu}_{H^+}$ (Section
7.2). As with other systems, there is dissent over the H^+/ATP stoicheiometry,
values of 2 and 3 being both quoted (see Hauska & Trebst 1977).

6.6 BACTERIORHODOPSIN AND THE PURPLE MEMBRANE OF HALOBACTERIA

Reviews Henderson & Unwin 1975, Stoekenius 1976, Stoekenius *et al.* 1979, Eisenbach & Caplan 1977, 1979

Halobacteria are extreme halophiles, requiring very high concentrations of NaCl and Mg^{2+} salts for growth, and they can therefore colonize environments such as salt lakes. *Halobacterium halobium* has been most studied. When grown under aerobic conditions the bacterium utilizes a conventional respiratory chain; however, when growing in the light under conditions of low oxygen tension they synthesize purple patches on their membrane which may be isolated by decreasing the osmolarity of the medium; the purple patches remain intact while the remainder of the cell membrane disintegrates. This "purple membrane" consists of flat sheets containing a hexagonal crystalline array of a single protein, bacteriorhodopsin, which makes up about 75% of the membrane dry weight, the remainder being phospholipid. The protein consists of a single polypeptide with a molecular weight of 26 000. The colour is due to a retinal molecule which is covalently bound as a Schiff base to a lysine side-chain (Fig. 6.12). The two-dimensional array of bacteriorhodopsin in the purple membrane has enabled very sophisticated image reconstruction techniques to be applied to electron micrographs of the membrane, with the result that the 3-D structure has been established at a

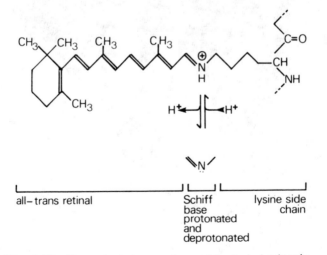

Fig. 6.12 The retinal chromophore of bacteriorhodopsin.

resolution of 7 Å (Unwin & Henderson 1975, Henderson & Unwin 1975). The protein is folded into 7 a-helical regions, each of which spans the membrane (Fig. 6.13).

Although *H. halobium* lacks chlorophyll, the cells can use light to eject protons and synthesize ATP. It can easily be shown that H^+-translocation is

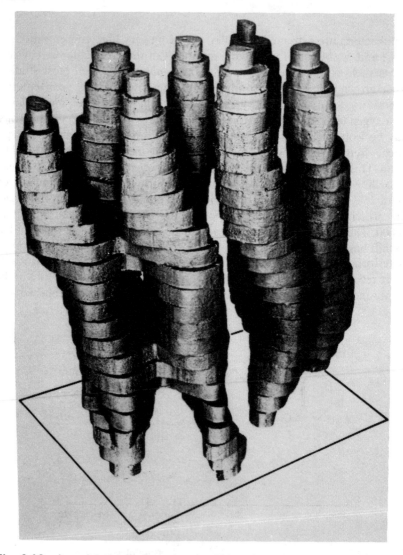

Fig. 6.13   A model of a single molecule of bacteriorhodopsin, viewed parallel to the membrane. (After Henderson & Unwin 1975.)

due to bacteriorhodopsin, since the purple membranes can be reconstituted into closed vesicles by the addition of phospholipid. These vesicles, which are inverted with respect to the intact cell, take up protons on illumination. Racker & Stoekenius (1974) added beef-heart ATP-synthetase to these vesicles and induced light-dependent ATP-synthesis. The importance of this demonstration was that it was difficult to explain in terms of any direct coupling mechanism, since bacteriorhodopsin and beef-heart ATP-synthetase had never met until that moment and so were unlikely to be able to be capable of direct interaction.

Bacteriorhodopsin is the simplest known proton pump. It differs from other light-driven or respiratory proton pumps in that H^+-translocation is not associated with electron transfer. The photochemical reactions of the pigment are complex (Fig. 6.14) and not yet fully solved. On illumination the pigment is bleached, and there is a vectorial release of a proton from the extracellular face of the membrane. The bleached form of the pigment, designated "M", then regenerates the purple pigment with the uptake of a proton from within the cell. A number of additional spectral intermediates have been resolved by low-temperature and laser-flash spectrometry. The only group which has been so far shown to undergo reversible protonation and deprotonation is the Schiff base linking retinal to the protein (Fig. 6.12). This is protonated in the original pigment but deprotonated in the "M"-form.

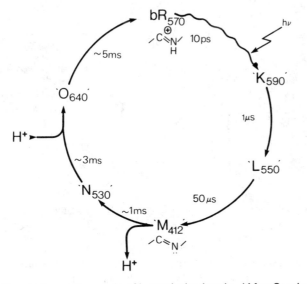

Fig. 6.14 The photoreaction cycle of bacteriorhodopsin. (After Stoekenius *et al.* 1979.)

Racker assembles the cold-labile F_1–ATP synthetase (negatively stained with phosphotungstate), while Mitchell juggles protons and charges, Slater attempts to grasp the elusive squiggle, and Boyer induces a conformational strain

7 The ATP Synthetase

7.1 INTRODUCTION

The ATP synthetase is a universal feature of energy-transducing membranes. It is present in mitochondria, chloroplasts, and both aerobic and photosynthetic bacteria, even including those bacteria which lack a functional respiratory chain and rely on glycolysis (Section 4.7). The structure of the complex is very similar in each of these membranes and distinct from that of other ATP-hydrolysing ion pumps, such as the $(Na^+ + K^+)$-stimulated ATPase of eukaryotic plasma membranes or the Ca^{2+}-ATPase responsible for accumulating Ca^{2+} within sarcoplasmic reticulum. In each case the function of the ATP-synthetase is the same—to utilize $\Delta\tilde{\mu}_{H+}$ to maintain the mass–action ratio for the ATPase reaction at least seven orders of magnitude away from equilibrium, or in the case of fermentative bacteria, to utilize ATP to maintain $\Delta\tilde{\mu}_{H+}$ for the purpose of transport.

7.2 THE STRUCTURE OF THE ATP SYNTHETASE

Reviews Senior 1973, Pedersen 1975, Racker 1975, 1976, Penefsky 1979 (mitochondria); Nelson 1976, Junge 1977, Shavit 1980 (chloroplasts); Kagawa 1978, Downie *et al.* 1979 (prokaryotes)

The ATP synthetase can be seen under the electron microscope in preparations of energy-transducing membranes which have been negatively stained with phosphotungstate. The complexes appear as knobs or mushrooms projecting from one side of the membrane. The orientation of the knobs is characteristic: in bacteria and mitochondria they project into the matrix or internal space; in isolated thylakoid membranes, chromatophores, or sonicated sub-mitochondrial particles they project outwards. In either case the structural and functional orientations correspond: ATP is hydrolysed or synthesized on the side of the membrane from which the knobs project, while protons cross from the side which lacks knobs during ATP synthesis (Fig. 1.1).

The ATP synthetase complex contains at least nine distinct polypeptides, some of which are present in multiple copies. Understanding the role of each of these components is the central problem in this field. The complex can be considered in two halves: the "knob" seen under the electron microscope corresponds to the catalytic site of ATP synthesis and is termed F_1 in mitochondria, CF_1 in chloroplasts, and TF_1 in thermophilic bacteria; the remainder of the ATP synthetase consists of hydrophobic proteins which are buried in the membrane and appear to be responsible for the conduction of the protons through the membrane towards F_1. This hydrophobic portion is termed F_0.

F_1 can be detached from F_0, and hence from the membrane, by a variety of treatments including urea, chelating agents or low ionic strength. The separated F_1 is soluble and has a molecular weight of about 360 k. It can be dissociated into 5 (sometimes 6) different polypeptides (Fig. 7.1). The sub-unit composition is very similar in preparations from different sources, and typical apparent molecular wights calculated from SDS-polyacrylamide electrophoresis (Fig. 7.1) are 56 kD (a-subunit), 53 kD (β-subunit), 33 kD (γ-subunit), 16 kD (δ-subunit) and 11 kD (ε-subunit). There is controversy over how many copies of each subunit are present, compositions of either $a_2\beta_2\gamma_2\delta_{1-2}\varepsilon_2$ or $a_3\beta_3\gamma\delta\varepsilon$ have been proposed.

The properties of F_1 alter when it is removed from the membrane. For example, except in the case of the extremely stable TF_1 from the thermophilic bacterium PS3 (see Kagawa et al. 1979) isolated F_1 is cold labile, dissociating reversibly into its individual subunits with loss of catalytic activity. Soluble F_1 catalyses the extremely rapid hydrolysis of ATP. At 25°C one mole of F_1 can hydrolyse 10^4 mole of ATP per min. In contrast the reverse reaction, ATP synthesis, can never be observed with soluble F_1. The nucleotide specificity of soluble F_1 is not absolute, and ITP and GTP can also be hydrolysed.

F_1 can be further dissected without losing catalytic activity. Thus CF_1 on treatment with trypsin yields a complex which contains only a and β subunits but nevertheless retains some catalytic activity. It is probable that the catalytic site is associated with the β-subunit, as photoaffinity analogues of ATP, which bind covalently upon ultraviolet irradiation, are mainly associated with the β-subunit (Scheurich et al. 1978). The minor δ-subunit appears to be required for the binding of F_1 to F_0. There is a natural inhibitor of the synthetase which prevents ATP hydrolysis when $\Delta\tilde{\mu}_{H+}$ is low and has been termed the inhibitor protein. CF_1 is subject to a more potent inhibition than the mitochondrial enzyme.

The functions of the individual subunits of the ATP synthetase of E. coli

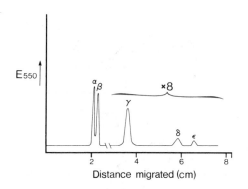

Fig. 7.1 The polypeptides of mitochondrial F_1-ATPase. (The γ, δ, and ε bands are enlarged eight-fold.)

have been investigated by examining "*unc*" mutants which possess defects in ATP synthesis (see Downie *et al.* 1979, Gibson *et al.* 1979). Five genes, *unc* A to E, have been distinguished which form part of an operon on the *E. coli* chromosome. The *unc* A and *unc* D genes have been shown to code for the *a* and *β* subunits of F_1 respectively; *unc* B and *unc* E genes probably code for components of F_0.

The complete F_1 does not bind directly to F_0 unless an additional component is present, the "oligomycin-sensitivity-conferring protein" or OSCP (18 kD molecular weight). Oligomycin is an antibiotic produced by *Streptomyces* which inhibits the ATP synthetase by binding to a component of F_0 (not to OSCP itself). Oligomycin interferes with the passage of protons through F_0 (Section 5.3). In the assembled $F_1.F_0$ this is sufficient to prevent both ATP synthesis and hydrolysis; oligomycin sensitivity is therefore diagnostic of a correct assembly of F_1 to F_0. Oligomycin is only effective with the mitochondrial enzyme, as OSCP is lacking in bacteria or chloroplasts. Dicyclohexylcarbodiimide (DCCD) has a different site of action but is effective also in bacterial and photosynthetic systems.

As with most membrane-bound proteins, the group of peptides which comprise F_0 are difficult to purify. The peptides are highly hydrophobic, and the complete $F_1.F_0$ complex can only be solubilized by treatments which destroy the membrane. Having solubilized the complex, F_1 and F_0 can be dissociated and separated.

The F_0 from the thermophilic bacterium PS3 has a molecular weight of 92 kD and contains three types of subunit: 6 copies of a 5·4 kD DCCD-binding protein, 3 copies of a 13 kD TF_1-binding protein and one 19 kD basic protein similar to the mitochondrial OSCP (Kagawa *et al.* 1979). The entire $F_1.F_0$ complex may be assembled as shown in Fig. 7.2.

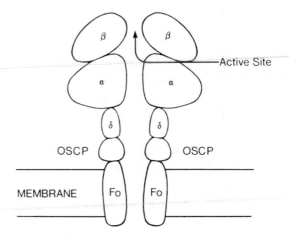

Fig. 7.2   The possible organization of the polypeptides of $F_1.F_0$. Adapted from Racker (1976). The scheme is hypothetical and the γ and ϵ subunits are omitted because their location is unknown.

7.3 THE FUNCTION OF F_0

Reviews Kagawa *et al.* 1979, Fillingame 1980

F_1 can easily be detached from inverted sub-mitochondrial particles. The resulting F_1-deficient particles lack not only the ability to synthesize or hydrolyse ATP, but are also inefficient in all energy-dependent processes. The addition of oligomycin, or the re-binding of F_1 restores the energy-transducing efficiency. As the effect of removing F_1 is similar to that of adding a proton translocator, this suggests that F_0 functions as a proton conductor to deliver protons across the membrane to the catalytic site of F_1. Normally proton conduction would be linked to ATP synthesis, but removal of F_1 would leave an uncontrolled proton conductor which was independent of ATP synthesis although still inhibitable by oligomycin or DCCD (Fig. 7.3).

This suggestion was confirmed when it was found that partially purified F_0 from beef heart mitochondria (Kagawa *et al.* 1973) or highly purified TF_0 from thermophilic bacteria (Okamoto *et al.* 1977) were shown to conduct protons in a reconstituted membrane (Fig. 7.4).

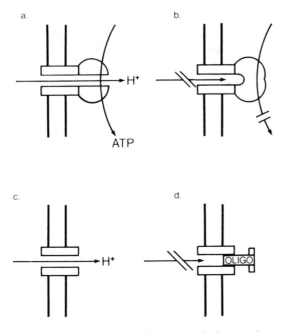

Fig. 7.3 Conditions for proton translocation through the membrane in intact and F_1-depleted organelles. The intact ATP synthetase only conducts protons when ATP is being synthesized (*a*). Proton translocation ceases when net ATP synthesis stops (*b*). In F_1 depleted ATP synthetase (*c*) an unregulated proton conductance exists, although it is still capable of inhibition by reagents which act on F_o, e.g. oligomycin (*d*).

7.4 THE MECHANISM OF ATP SYNTHESIS BY F_1

Reviews Mitchell 1974b, Boyer *et al.* 1977, Kozlov & Shulachev 1977

It is not known how the ATP-synthetase uses the proton electrochemical potential to alter the equilibrium of the ATPase reaction in the direction of ATP synthesis. The mechanism appears to be different from that of two other ATP-hydrolysing ion pumps, the $(Na^+ + K^+)$ ATPase of mammalian plasma membranes and the Ca^{2+}-ATPase of sarcoplasmic reticulum, since unlike these examples no phosphorylated intermediate has been detected in the ATP synthetase.

Mitchell (see Mitchell, 1974b) has proposed a direct mechanism in which protons are involved in ADP esterification at the catalytic site (Fig. 7.5). In

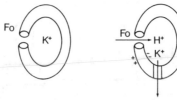

(a)   (b)

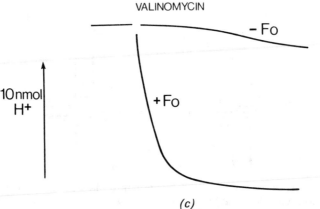

(c)

Fig. 7.4 Purified F_o conducts protons across artificial phospholipid bilayers. Highly purified F_o from the thermophilic bacterium PS3 was suspended together with soybean phospholipids in a cholate solution containing tricine buffer. The cholate was dialysed away for 16 hours so that vesicles formed (Section 1.3) with F_o incorporated into the bilayer. The vesicles were first loaded with KCl by incubating them at 55°C in a KCl medium. They were then washed and resuspended in a K^+-free medium in a chamber with a pH-electrode (a). Valinomycin was then added, and a diffusion potential was generated by K^+ efflux down its concentration gradient H^+-uptake through F_o in response to this potential was monitored (b). The phospholipid bilayer itself was highly impermeable to protons, since little H^+-uptake occurred in the absence of F_o (c). (Data from Okamotu *et al.* 1977.)

this model, two protons driven through F_0 by $\Delta\tilde{\mu}_{H+}$ attack one of the oxygen atoms of Pi, forming water and leaving an extremely reactive species which reacts directly with ADP to form ATP. In an alternative mechanism advanced by Boyer, Slater and others a less direct role for the translocated protons is proposed but more emphasis is placed on the polypeptides of F_1, suggesting that the Gibbs energy of the proton gradient is used to induce conformational changes in F_1 (Slater 1974, Boyer 1977). The conformational

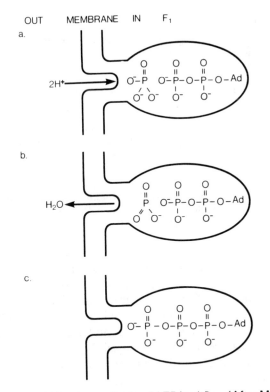

Fig. 7.5 A direct model for the synthesis of ATP by $\Delta\tilde{\mu}_{H+}$. (After Mitchell 1974b.)

changes would in turn cause the synthesis of ATP by altering the relative affinities of F_1 for the substrates and products of the reaction. The actual formation of ATP at the catalytic site would involve rather small energy changes, whereas the ultimate release of bound ATP from F_1 would be the major energy-requiring step. There is a precedent for such a mechanism in the hydrolysis of ATP by myosin during muscle contraction (Fig. 7.6) where binding of ATP involves large Gibbs energy changes and causes large conformational changes in the protein, while hydrolysis of the bound ATP causes much smaller Gibbs energy changes (see Chappell 1977).

Three lines of evidence support this model. First $\Delta\tilde{\mu}_{H+}$ alters the conformation of chloroplast CF_1. Experiments by Ryrie & Jagendorf (1972) showed that the slow exchange of ^3H from ^3H_2O into the hydrogen atoms of CF_1 was increased when light was used to generate a $\Delta\tilde{\mu}_{H+}$, suggesting an alteration in the tertiary structure of the enzyme. N-ethylmaleimde also showed an increased binding to the γ-subunit of CF_1 in the light (see McCarty 1979). This

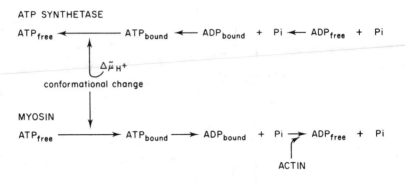

Fig. 7.6 The indirect model for the ATP synthetase has features in common with the mechanism of ATP hydrolysis by the muscle actomyosin complex. (After Chappell 1977.)

conformational change affects the catalytic activity of CF_1. The enzyme is latent in the chloroplast unless a $\Delta\tilde{\mu}_{H+}$ exists across the membrane and will therefore not hydrolyse ATP in the presence of a proton translocator. Following the $\Delta\tilde{\mu}_{H+}$-induced conformational change, thiol reagents or trypsin modify CF_1 so that its catalytic activity is no longer inhibited at low $\Delta\tilde{\mu}_{H+}$. Conformational changes in mitochondrial F_1 have been deduced from changes in the fluorescence of the inhibitor aureovertin, which binds to F_1 (see Slater 1974).

F_1 and CF_1 both possess tight binding sites for adenine nucleotides, although opinion is divided as to whether the nucleotides bound to these sites are on the main path of ATP synthesis or whether their role is regulatory (McCarty, 1979).

The third approach has come from isotope exchange studies. This technique has the advantage that it can detect the formation of minute amounts of product, for example enzyme-bound intermediates. The most informative exchange is that between $H_2^{18}O$ and Pi. This is believed to represent the dynamic equilibrium of ATP synthesis at the catalytic site without requiring that the bound ATP be released into solution (Fig. 7.7). This exchange is inhibited by oligomycin and requires ADP but can still occur at low $\Delta\tilde{\mu}_{H+}$, suggesting that ATP synthesis is not the major energy-requiring step, but rather it is ATP release. A more detailed analysis of the exchange reactions suggests that F_1 contains two catalytic sites which function alternately (Boyer 1977).

A change in the conformation of F_1 does not in itself explain the coupling of ATP synthesis to proton translocation. As with conformational models of proton translocation by electron-transfer chains (Section 5.4), a conforma-

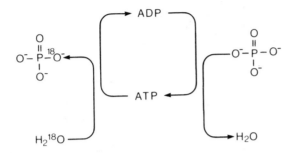

Fig. 7.7 The incorporation of oxygen-18 from water into Pi is indicative of the reversible synthesis of ATP bound to the ATP synthetase. (Note that the oxygen bridging the β and γ phosphates comes from ADP and not Pi.)

tional model for the ATP synthetase must co-ordinate a number of steps. A simple model is shown in Fig. 7.8, in which the energy inherent in the proton electrochemical potential is extracted by altering the binding affinities for the proton-binding site such that it has a low affinity when facing the phase of high electrochemical potential (the C-phase) and a high affinity for protons when facing the phase of low proton electrochemical potential (the M-phase).

7.5 THE TRANSPORT OF ADENINE NUCLEOTIDES AND Pi IN MITOCHONDRIA

Reviews Fonyo 1975 (phosphate carrier); Vignais 1976, Klingenberg 1979a (adenine nucleotide translocator)

In bacteria and chloroplasts the ATP synthetase produces ATP in the same compartment in which it is utilized, but mitochondria synthesize ATP in the matrix and then export the nucleotide to the cytosol. Two carriers are involved: the phosphate carrier for the uptake of Pi, and the adenine nucleotide translocator for the uptake of ADP and export of ATP (see Fig. 7.12).

The phosphate carrier catalyses the electroneutral transport of $H_2PO_4^-$, either in exchange for OH^- or by symport with a proton, the two being indistinguishable. The carrier is inhibited by mercurial reagents such as p-mercuribenzoate and mersalyl, and also by N-ethylmaleimide, although none of the inhibitors is completely specific. The phosphate carrier is extremely active. Because of the proton symport, the distribution of Pi across the membrane is influenced by ΔpH, and a factor complicating the measurement of H^+/O ratios in mitochondria by oxygen pulse methods is the facility

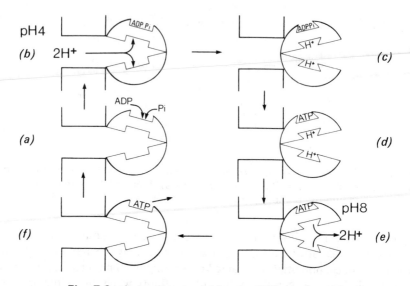

Fig. 7.8 A possible model for the ATP synthetase.
F_1 has two conformations, a deprotonated form, characterized by low-affinity proton-binding sites facing F_0 together with a low-affinity catalytic site, and a protonated form in which high-affinity proton-binding sites (i.e. with an increased pK) face away from F_0. This conformation also results in an increased affinity of the catalytic site. (a) ADP and Pi bind to the low-affinity form of the catalytic site. (b) Protons from F_0 bind to the low-affinity proton-binding sites; this causes a conformational change (c) as a result of which the affinity of the catalytic site greatly increases. The free energy for this transformation is derived from a corresponding increase in the affinity with which the protons are bound. (d) ATP tightly bound to the catalytic is now produced by a step which involves relatively little free-energy change. (e) If the activity of protons in the right-hand phase is sufficiently low, protons can dissociate from their binding sites despite the high pK in this conformational state. (f) When the protons dissociate, the conformation reverts to its original form, and the catalytic site reverts to the low-affinity conformation, thus allowing release of the bound ATP.

with which Pi can redistribute across the membrane and partially neutralize any ΔpH generated by respiration (Section 4.3).

The adenine nucleotide translocator catalyses the 1:1 exchange of adenine nucleoside di- or triphosphates across the inner membrane (Fig. 7.9). The total pool size of adenine nucleotides in the matrix (ATP + ADP + AMP) does not change, as the uptake of a cytosolic nucleotide is automatically compensated by the efflux of a nucleotide. Even if mitochondria are suspended in a nucleotide-free medium (as they are during preparation), no loss of nucleotide normally occurs.

The translocator is specific for ATP and ADP (not AMP). A number of

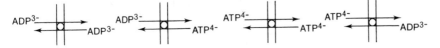

Fig. 7.9 Exchanges catalysed by the adenine nucleotide translocator.

inhibitors are specific for the translocator. Atractyloside, a glucoside isolated from the Mediterranean thistle *Atractylis gummifera*, is a competitive inhibitor of adenine nucleotide binding and transport. The closely related carboxyatractylate binds more firmly (K_d 10^{-8}M) and cannot be displaced by adenine nucleotides. Bongkrekic acid is produced by *Pseudomonas cocvenenans* and derives its name from its discovery as a toxin in contaminated samples of the coconut food product bongkrek. It is an uncompetitive inhibitor of the translocator.

In the model of the translocator proposed by Klingenberg (1979a), the carrier can exist in two states, which differ in the orientation of its single nucleotide binding site (Fig. 7.10). The C-state, in which the binding site is accessible from the cytosol is stabilized by carboxyatractylate, while the M-state is fixed by bongkrekic acid.

Two approaches have been used to identify the peptide in the inner membrane responsible for adenine nucleotide translocation. When [^{35}S]-carboxyatractylate is added to beef-heart mitochondria the binding is sufficiently stable to survive solubilization of the membrane with the non-ionic detergent Triton X100 and subsequent purification. The inhibitor enables the peptide to be identified and protects it against inactivation during purification. Virtually all the isotope is found to be associated with a 40 kD protein. A second approach has come from the use of photoaffinity analogues of adenine nucleotides, which are competitive inhibitors of transport which bind to the catalytic site of the carrier in the dark. In ultraviolet light they lose N_2 to form highly reactive nitrene-free radicals which bind covalently to the nearest peptide, i.e. the translocator.

The binding protein has an apparent molecular weight of 30 kD. From the amount of labels bound it is likely that the intact translocator acts as a dimer of 60 kD. In heart the translocator is the most abundant protein on the mitochondrial inner membrane.

While the translocator transports ADP and ATP symmetrically when there is no membrane potential, under normal respiring conditions uptake of ADP and efflux of ATP are preferred, corresponding to the physiological direction of the exchange. The reason for this asymmetry lies with the relative charges on the two nucleotides. ATP is transported as ATP^{4-}; ADP is transported as ADP^{3-}. The resulting charge imbalance means that the

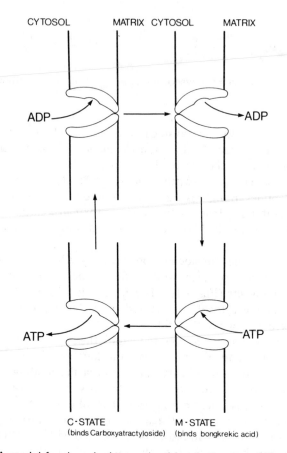

Fig. 7.10   A model for the adenine nucleotide translocator. (After Klingenberg 1979a.)

equilibrium of the exchange is affected tenfold for each 60 mV of membrane potential (Fig. 7.11).

The complete system for mitochondrial ATP synthesis and export can now be re-assembled (Fig. 7.12), with the ATP synthetase, adenine nucleotide translocator and phosphate carrier. The combined effect of the phosphate carrier and the adenine nucleotide translocator is to cause the influx of one additional proton per ATP synthesized. Note that although the additional proton apparently enters with Pi this is an electroneutral process and the charge of the additional proton is used to drive the exchange of ADP^{3-} for ATP^{4-}

The thermodynamic consequences of this are considerable. First, up to

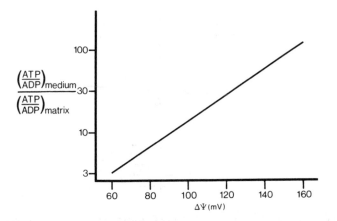

Fig. 7.11 Membrane potential affects the equilibrium of the adenine nucleotide translocator.

If there is a net charge imbalance of one when ADP exchanges for ATP, it would be predicted (Section 3.5) that the ATP/ADP ratio inside the matrix would be ten-fold lower than in the medium for each 60 mV of membrane potential, under equilibrium conditions. This was examined by Klingenberg and Rottenberg (1977). Mitochondria were incubated in a medium containing valinomycin so that $\Delta\Psi$ could be determined from the distribution of $^{86}Rb^-$ across the membrane (Section 4.2). The incubation also contained $[^{14}C]ATP$ and $[^{14}C]ADP$, oligomycin to inhibit the ATP synthetase (Section 7.2), and succinate as respiratory substrate. $\Delta\Psi$ was varied from 60–170 mV by altering the concentration of K^+ in the medium from 20 mM to 0.5 mM. After allowing enough time for the adenine nucleotides to distribute to equilibrium across the inner membrane, the mitochondria were separated from the incubation medium by centrifugation through silicone oil (Fig. 4.5) and the ATP and ADP contents of the mitochondria and medium were determined. The disparity between the ATP/ADP ratios in the medium and matrix increased with $\Delta\Psi$, until when $\Delta\Psi$ was 160 mV ATP/ADP in the medium was 125 times higher than in the matrix. (Data from Klingenberg & Rottenberg 1977.)

one-third of the Gibbs energy of the cytosolic ATP/ADP + Pi pool comes, not from the ATP synthetase itself, but from the subsequent transport. Secondly, as three protons are used to synthesize a cytosolic ATP but only two for a matrix ATP, it follows that in State 4 the cytosolic ΔG_p (Section 3.2) can be up to 50% higher than in the matrix, or that produced by inverted sub-mitochondrial particles. This is seen; isolated mitochondria can maintain a ΔG_p of up to 64 kJ mole^{-1} (Slater *et al.* 1973) in contrast to a value of less than 50 kJ mole^{-1} for sub-mitochondrial particles. A requirement for 3 protons for each ATP has also consequences for proposed mechanisms of proton extrusion by the respiratory chain (Section 4.3).

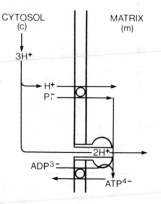

Fig. 7.12 The translocation of ADP and Pi into the matrix and the export of ATP involves the translocation of an additional proton.
(For simplicity the Pi^-/OH^- antiport is expressed as the equivalent of H^+/Pi^- symport.)

ATP SYNTHETASE	$ADP^{3-}_{(m)} + Pi^-_{(m)} + 2H^+_{(c)} \rightarrow ATP^{4-}_{(m)} + 2H^+_{(m)}$
PHOSPHATE TRANSPORT	$Pi^-_{(c)} + H^+_{(c)} \rightarrow Pi^-_{(m)} + H^+_{(m)}$
TRANSLOCATOR	$ADP^{3-}_{(c)} + ATP^{4-}_{(m)} \rightarrow ADP^{3-}_{(m)} + ATP^{4-}_{(c)}$
OVERALL	$ADP^{3-}_{(c)} + Pi^-_{(c)} + 3H^+_{(c)} \rightarrow ATP^{4-}_{(c)} + 3H^+_{(m)}$

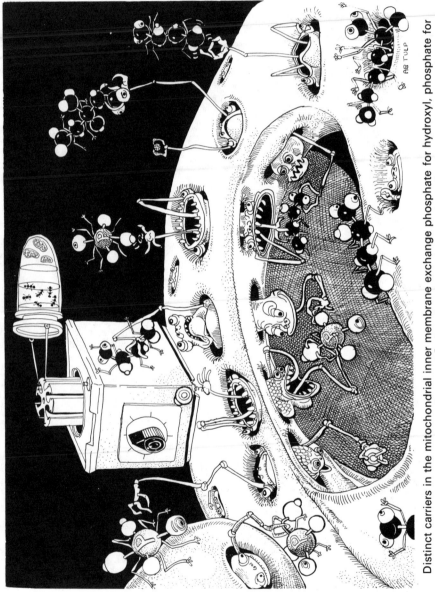

Distinct carriers in the mitochondrial inner membrane exchange phosphate for hydroxyl, phosphate for malate, and malate for citrate (plus a proton)

8 The Interaction of Bioenergetic Organelles with Their Environment

8.1 INTRODUCTION

Bioenergetic organelles can only function if there is a continual interchange of metabolites and end products with the cell cytosol (in the case of mitochondria and chloroplasts) or with the external environment (in the case of bacteria). In addition the organelles must regulate their ionic composition while at the same time maintaining a high $\Delta\tilde{\mu}_{H^+}$ for ATP synthesis. To perform these functions simultaneously, an elaborate system of ion- and metabolite-carriers has evolved, and these will be considered in this chapter.

8.2 METHODS FOR THE STUDY OF METABOLITE TRANSPORT

Reviews Palmieri & Klingenberg 1979, LaNoue & Schoolwerth 1979

This section will be concerned primarily with the methodology for investigating mitochondrial metabolite transport, although the techniques are equally applicable to other systems.

The combination of high transport rates and small internal volumes poses a considerable experimental problem. While most transport processes occurring across the plasma membrane of eukaryotic cells have $t_{1/2}$ values of several hours, transport across the mitochondrial inner membrane — be it net flux or exchange — results in transients which are complete within a few seconds.

A rapid qualitative technique for detecting metabolite transport relies on the osmotic swelling of the matrix induces by bulk solute entry (Section 2.7). If non-respiring mitochondria are suspended in an iso-osmotic salt solution of the anion in question, swelling, and consequently decreased light-scattering, will occur if there is a pathway for both cation and anion and if charge and pH balance is retained (Section 2.7). Metabolite entry is often linked

167

ultimately to proton symport, and in these cases matrix acidification can be avoided by the use of ammonium salts, NH_3 being the permeant species (Section 2.4) protonating in the matrix to form NH_4^+.

The "ammonium swelling" technique was developed by Chappell (1968) to detect electroneutral transport systems. Phosphate transport (Section 7.6) proved to be equivalent to a $H^+/H_2PO_4^-$ symport Fig. 8.1, since rapid swelling occurred in ammonium phosphate. In contrast, swelling in ammonium malate only occurred in the presence of a low concentration of Pi, suggesting a linkage between the phosphate carrier and a malate/phosphate exchange.

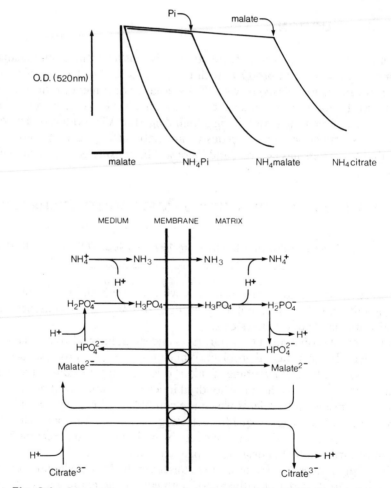

Fig. 8.1 Ammonium-swelling and mitochondrial metabolite carriers.

Swelling in ammonium citrate required the presence of both Pi and malate, suggesting in this case a three-stage cascade.

The swelling technique is difficult to quantitate and requires high (iso-osmotic) concentrations of anions. A more direct technique which avoids these problems follows the exchange or net transport of radio-labelled metabolites into the mitochondrial matrix. The rapidity with which the matrix metabolite pools equilibrate with the incubation necessitates the use of rapid techniques if kinetics are to be resolved. Silicone-oil centrifugation (Fig. 4.5) is most frequently employed, and the time-resolution may be improved by terminating transport prior to centrifugation by the addition of a specific transport inhibitor. With this inhibitor-stop method, the time resolution may be improved to less than one second. Automated devices have been developed which allow sequential samples to be taken from a single incubation, thus enabling the kinetics to be determined.

Most mitochondrial transport systems catalyse exchange reactions, so it is essential to load the mitochondria with the appropriate exchange partner to obtain reliable results.

8.3 MITOCHONDRIAL METABOLITE CARRIERS

Reviews Meijer & van Dam 1974, LaNoue & Schoolwerth 1979

Table 8.1 lists the major metabolite carriers which have now been identified in the mitochondrial inner membrane; the list may not be exhaustive. The adenine nucleotide translocator and the phosphate carrier have already been considered in the context of ATP synthesis (Section 7.6). As oxidative phosphorylation is a universal mitochondrial function, these two carriers are ubiquitous. Similarly, as pyruvate and fatty acids are major mitochondrial substrates *in vivo*, the pyruvate and carnitine carriers are also widely distributed.

The firm identification of the pyruvate carrier is relatively recent. Because pyruvate is a monocarboxylic acid, it had been argued that it could cross bilayer regions without the need for a carrier, following the precedent of acetate (Section 2.4). However, the demonstration of saturation kinetics and of a specific inhibitor, cyanohydroxycinnamate, confirmed the involvement of a carrier. As with the other carriers, it should be noted that pyruvate $^-$/OH$^-$ antiport is indistinguishable from pyruvate $^-$/H$^+$ symport.

The carnitine carrier has also been identified recently. Previously it was considered that acylcarnitine entry into the matrix was an example of group

Table 8.1   Mitochondrial metabolite carriers

Carrier	Function	Inhibitors
(a) Adenine nucleotide translocator	$ADP^{3-} \rightarrow$ / $\leftarrow ATP^{4-}$	Atractylate Carboxyatractylate Bongkrekate
(b) Phosphate carrier	$Pi^{-} \rightarrow$ / $\leftarrow OH^{-}$	Sulphydryl reagents (*N*-ethyl maleimide) (mersalyl)
(c) Dicarboxylate	$malate^{2-} \rightarrow$ / $\leftarrow Pi^{2-}$	Butylmalonate
(d) Tricarboxylate	$citrate^{3-} + H^{+} \rightarrow$ / $\leftarrow malate^{2-}$	1,2,3-Benzyl-tricarboxylate
(e) 2-oxoglutarate	$2\text{-}oxoglut^{2-} \rightarrow$ / $\leftarrow malate^{2-}$	Phenylsuccinate
(f) Glutamate–aspartate	$glu^{2-} + H^{+} \rightarrow$ / $\leftarrow asp^{2-}$	———
(g) Glutamate	$glu^{-} \rightarrow$ / $\leftarrow OH^{-}$	
(h) Pyruvate	$pyr^{-} \rightarrow$ / $\leftarrow OH^{-}$	Cyanohydroxycinnamate
(i) Carnitine	$acylcarnitine^{+} \rightarrow$ / $\leftarrow carnitine^{+}$	
(j) Ornithine	$ornithine \rightarrow$ / $\leftarrow H^{+}$	

translocation (Section 1.4) in which only the acyl group, and not the carnitine, actually crossed the membrane. The carnitine carrier will exchange carnitine for acetylcarnitine, short- or long-chain acylcarnitines.

Two carriers which also occur in many different mitochondria are the 2-oxoglutarate (i.e. *a*-ketoglutarate) carrier, and the glutamate–aspartate carrier (Table 8.1). These together are involved in the malate–aspartate shuttle (Fig. 8.2), a device allowing the oxidation of cytosolic NADH by the respiratory chain, despite the fact that the inner membrane is impermeable to the nucleotide. An additional problem posed by this process is that the E_h of the cytosolic $NAD^+/NADH$ couple is considerably higher (i.e. less reducing) than the equivalent matrix couple. This thermodynamic impasse is overcome by the electrical inbalance of the glutamate–aspartate carrier, which exchanges glutamate^{2-} plus a proton for aspartate^{2-} and is therefore driven in the direction of glutamate uptake and aspartate expulsion when a $\Delta\tilde{\mu}_{H^+}$ exists across the membrane.

The a-glycerophosphate shuttle provides a second means for the oxidation of cytosolic NADH (Fig. 8.3). This makes use of the two a-glycerophosphate dehydrogenases present in most cells—a cytosolic enzyme coupled to NAD^+ and an enzyme on the outer face of the inner mitochondrial membrane feeding electrons directly to UQ. In this case the directionality is induced by feeding electrons to the quinone pool at a potential close to zero millivolts (Fig. 5.11).

The dicarboxylate and tricarboxylate carriers are highly active in liver

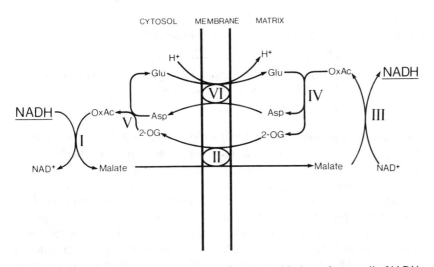

Fig. 8.2 The glutamate–aspartate cycle for the oxidation of cytosolic NADH. (I) Cytosolic NADH oxidized by cytosolic malate dehydrogenase; (II) Malate enters matrix in exchange for 2-oxoglutarate; (III) Malate re-oxidized to oxaloacetate by matrix malate dehydrogenase, genereating matrix NADH; (IV) matrix oxaloacetate transaminates with glutamate to form aspartate and 2-oxoglutarate, which exchanges out of the matrix; (V) 2-oxoglutarate transaminates in cytosol with transported aspartate to regenerate cytosolic oxaloacetate and gives cytosolic glutamate, which re-enters the matrix (VI) by proton symport in exchange with aspartate.

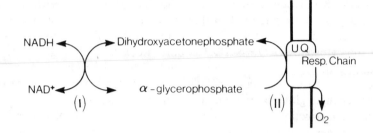

Fig. 8.3 The α-glycerophosphate shuttle for the oxidation of cytosolic NADH. (I) cytosolic NAD⁺-linked dehydrogenase; (II) mitochondrial enzyme.

mitochondria but almost absent from heart. Both allow the net export of citric acid cycle intermediates from the matrix to be used for gluconeogenesis and fatty acid synthesis respectively. Both these carriers are electroneutral, the tricarboxylate carrier accomplishing this by co-transporting a proton together with the citrate.

8.4 MITOCHONDRIAL CALCIUM TRANSPORT

Reviews Bygrave 1977, Carafoli & Crompton 1977, Nicholls & Crompton 1980, Saris & Åkerman 1980

The chemiosmotic theory provided a simple observation for the enormous capacity of mitochondria to accumulate Ca^{2+}, since the electrical uniport of a divalent cation across a membrane sustaining around 180 mV of membrane potential would be predicted to result in an equilibrium concentration gradient of no less than 10^6 (Fig. 3.5). However, the very magnitude of this accumulation presents problems in a cellular context, since the presence of Pi within the matrix probably prevents the free matrix Ca^{2+} concentration from rising much above 1 mM, implying that the equilibrium cytosolic free Ca^{2+} concentration should be as low as 10^{-9} M. In contrast most estimates of cytosolic free Ca^{2+} concentration are in the range 10^{-7} to 10^{-6} M. Not only is a single uniport carrier too powerful, but also it would only enable the distribution of Ca^{2+} between cytosol and matrix to be regulated by altering the membrane potential (Eq. 3.29), which would in turn disturb ATP synthesis and metabolite distribution.

The solution to this dilemma came with the discovery of independently operating efflux pathways in liver (Puskin *et al.* 1976) and heart (Crompton &

Carafoli 1976) mitochondria. In the case of heart, brain or brown fat mito-chondria the efflux pathway exchanges Ca^{2+} for Na^+, while in the case of liver Ca^{2+} is ultimately exchanged for protons (Fig. 8.4). As the equilibrium distributions sought by the efflux pathways are quite different from that of the uniporter (Fig. 3.4), a slow continuous cycling of Ca^{2+} occurs across the

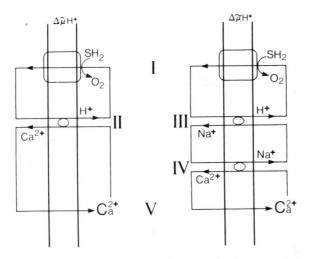

Fig. 8.4 Steady-state cycling of Ca^{2+} in liver (*left*) and heart (*right*) mitochondria. (I) Respiratory chain; (II) Ca^{2+}/H^+ exchange in liver mitochondria; (III) Na^+/H^+ exchange in heart mitochondria; (IV) Ca^{2+}/Na^+ exchange in heart mitochondria; (V) Ca^{2+} uniporter. Stoicheiometries are not specified.

membrane, at the expense of a slight utilization of $\Delta\tilde{\mu}_{H+}$ (Fig. 8.4). This two-pathway system has a number of advantages. Not only can the steady-state distributed be much less than 10^6, but the distribution can be regulated by altering the kinetic parameters of either pathway without altering $\Delta\Psi$ and ATP synthesis. Finally, as with any cycling system regulation can be very sensitive: a slight change in the rate of either pathway can induce a large percentage change in the net flux.

A wide range of methods exist to study mitochondrial Ca^{2+} transport. Besides the direct determination of Ca^{2+} by atomic absorption, isotopic measurements may be made with $^{45}Ca^{2+}$, Ca^{2+}-selective electrodes may be used, or metallochromic indicators such as arsenazo III, which undergo a spectral change on binding Ca^{2+}, can be employed (Scarpa 1979).

The existence of the independent efflux pathway can be demonstrated most simply by the selective inhibition of the uptake pathway once steady-state conditions have been obtained. One such inhibitor is the glycoprotein stain

Ruthenium Red, which inhibits the uniporter at very low concentrations. As the efflux pathway is unaffected, a net efflux of Ca^{2+} from the matrix is seen, (Fig. 8.5), even though $\Delta\Psi$ is too high for efflux to occur by reversal of the uniporter. The uniporter may also be inhibited by other polycations, such as Mg^{2+} and lanthanides (which also affect the efflux pathway at higher concentrations).

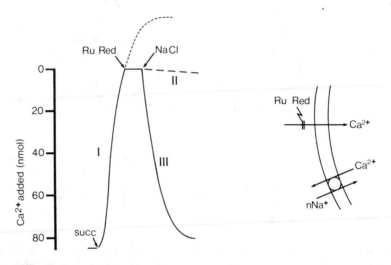

Fig. 8.5 Na^+-dependent efflux of Ca^{2+} from heart mitochondria.
Heart mitochondria were incubated in a KCl medium in the presence of 70 nmol of Ca^{2+} mg protein^{-1}. Succinate was added as substrate where shown, and the activity of the Ca^{2+}-uniporter led to a net uptake of Ca^{2+} (I); the efflux pathway was not active, owing to the absence of Na^+. The uniport inhibitor, Ruthenium Red was then added and net Ca^{2+} transport virtually ceased (II). 13mM NaCl was then added, activating the efflux pathway, as the uniporter was still inhibited, a complete efflux of accumulated Ca^{2+} occurred (III). (Data from Crompton *et al.* 1976.)

   The steady-state cycling of Ca^{2+}, H^+ and (in the case of heart) Na^+ across the membrane involves no net movement of ions across the membrane (Fig. 8.4). However, when *net* Ca^{2+} accumulation occurs, the entry of the cation must be compensated. In the strict absence of electroneutrally permeant weak acid, Ca^{2+} uptake is charge balanced by the extrusion of protons by the respiratory chain, with a stoicheiometry of $2H^+/Ca^{2+}$. This leads to a rapid increase in ΔpH across the membrane, with a consequent decrease in $\Delta\Psi$ ($\Delta\bar{\mu}_{H+}$ being roughly constant (Fig. 4.8). Under these conditions (e.g. in the presence of NEM to inhibit movement of endogenous Pi (Section 7.6)) Ca^{2+}

accumulation is very limited, and a ΔpH of -2 units can rapidly build up.

If an electroneutrally permeant weak acid is present, it can move into the matrix in response to this ΔpH, dissipating the pH gradient, and allowing $\Delta \Psi$ to increase, which in turn allows further Ca^{2+} uptake. Acetate is one such anion, while Pi not only dissipates ΔpH but forms a $Ca_3 (PO_4)_2$ complex in the matrix, allowing very large amounts of Ca^{2+} to be stored. For reasons which are not clear, mitochondria sometimes react adversely to the accumulation of Ca^{2+} in the presence of Pi, exhibiting swelling, membrane potential collapse and Ca^{2+} release. This "damage" can be potentiated by PEP, atractylate, or by the oxidation of matrix NAD(P)H (e.g. by AcAc), and is prevented by the presence in the medium of low concentrations of ATP or ADP. This "damage" efflux must be distinguished carefully from the independent efflux pathway described above which occurs without a collapse in $\Delta \Psi$.

8.5 BACTERIAL TRANSPORT

Reviews Hamilton 1975, 1977, Harold 1977, Wilson 1978

Bacteria are forced to survive in environments which are far more variable, and usually more hostile, than anything experienced by a mitochondrion or chloroplast. As a consequence they have developed a variety of mechanisms for the uptake of metabolites such as amino acids and sugars. As the origins of the chemiosmotic theory lay in Mitchell's desire to explain "active transport" in bacteria, it is significant that one of the bacterial transport mechanisms, the phosphotransferase system, represents the most thoroughly authenticated example of vectorial group translocation. The second class of transport mechanism is "chemiosmotic" and linked to the proton circuit; the third consists of a set of metabolite pumps linked directly to ATP hydrolysis.

The proposed chemiosmotic mechanisms are shown in Fig. 8.6. The simplest assumption is that cationic metabolites such as lysine are transported with their positive charge by a uniport mechanism, in which case their equilibrium distribution would be a function of the membrane potential. Uncharged metabolites such as isoleucine would permeate by proton symport, so that accumulation would be a function of the total $\Delta \tilde{\mu}_{H+}$, while anionic species, also permeating by proton symport, would equilibrate with the pH gradient. There is no *a priori* reason why other proton stoicheiometries should be excluded (see Rottenberg 1976).

The best documented chemiosmotic transport mechanism is the classic *lac*

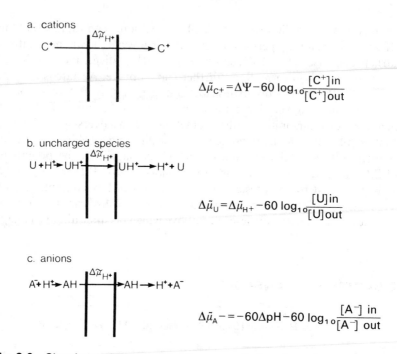

a. cations

$$\Delta\tilde{\mu}_{C+}=\Delta\Psi-60\,\log_{10}\frac{[C^+]in}{[C^+]out}$$

b. uncharged species

$$\Delta\tilde{\mu}_{U}=\Delta\tilde{\mu}_{H+}-60\,\log_{10}\frac{[U]in}{[U]out}$$

c. anions

$$\Delta\tilde{\mu}_{A}-=-60\Delta pH-60\,\log_{10}\frac{[A^-]\,in}{[A^-]\,out}$$

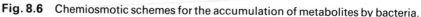

Fig. 8.6 Chemiosmotic schemes for the accumulation of metabolites by bacteria.

permease of *E. coli* (Fig. 1.13) which transports lactose and other β-galacto-sides (West & Mitchell 1972, 1973). Transport of the uncharged lactose, or its non-metabolizable analogue thiomethylgalactoside, is coupled to the sym-port of one proton.

ATP-dependent metabolite transport has been demonstrated in a variety of Gram-negative bacteria. Transport is associated with the presence, in the periplasmic space, of specific binding proteins, with molecular weights from 20–40 kD. These proteins are specific for one metabolite, which they bind with high affinity. Over 20 have been described. Osmotic shock leads to a loss of these proteins and to an impairment of ATP-dependent transport. The binding proteins do not represent the carriers *per se* but confer the affinity and specificity on the transport process. It is not yet clear whether there is a common transporter or whether each metabolite has its own.

Sensitivity to osmotic shock is one criterion which can be used to dis-tinguish between chemiosmotic and ATP-dependent mechanisms. In addi-tion, the former is sensitive to proton translocator, requires oxidizable substrate in mutants with a defective ATP synthetase, and is resistant to arsenate (which depletes ATP). In contrast "shockable" transport is resistant to proton translocators, is dependent on glycolysis not respiration in ATP-synthetase-deficient mutants, and is inhibited by arsenate. ATP-dependent transport also occurs in Gram-positive bacteria, but in these cases periplasmic binding proteins are absent.

The third transport mechanism, the phosphotranferase system, catalyses the transport of some sugars, including glucose, in bacteria such as *E. coli* and *Staphylococcus aureus*. The distinctive feature of this system is that the phosphorylation of transported sugars which preceeds their subsequent metabolism is here combined with the transport process to create a group translocation (Fig. 8.7). The mechanism is confined to anaerobic and facultative bacteria and is absent from aerobic species. All sugars are transported by the system in *S. aureus*, but only some, such as glucose enter by this mechanism in *E. coli*. Phosphoenolpyruvate acts as the phosphoryl donor, phosphorylating Enzyme I, which in turn phosphorylates a small heat-stable protein, HPr. Both these enzymes are soluble, constitutive, and do not show sugar specificity. In *S. aureus*, the phosphate group is transferred to a soluble Factor III, which is sugar specific. Factor III consists of three sub-units of molecular weight 12 kD, each of which can bind one phosphate. Finally, a ternary complex is formed between phosphorylated Factor III, the sugar, and a membrane bound, sugar-specific Enzyme II. The phosphate is then transferred from Factor III to the sugar as it is transported.

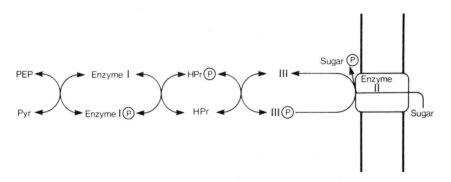

Fig. 8.7 The phosphotransferase system for sugar transport.

References

Åkerman, K. E. O. & Wikström, M. K. F. (1976). *FEBS Lett.* **68**, 191–197.

Avron, M. (1977). *Annu. Rev. Biochem.* **46**, 143–155.

Avron, M. (1978). *FEBS Lett.* **96**, 225–232.

Banks, B. E. C. & Vernon, C. A. (1978). *Trends Biochem. Sci.* **3**, N156–N158.

Bashford, C. L. & Smith, J. C. (1979). *Methods Enzymol.* **55**, 569–586.

Beinert, H. (1978). *Methods Enzymol.* **54**, 133–150.

Bennett, J. (1979). *Trends Biochem. Sci.* **4**, 268–271.

Blankenship, R. E. & Parson, W. W. (1978). *Annu. Rev. Biochem.* **47**, 635–653.

Blok, M. C., Hellingwerf, K. J. & Van Dam, K. (1977). *FEBS Lett.* **76**, 45–50.

Boyer, P. D. (1965) *In* "Oxidases and Related Redox Systems" (T. E. King, H. S. Mason & M. Morrison, ed.) pp. 994–1008. New York, Wiley.

Boyer, P. D. (1977). *Trends Biochem. Sci.* **2**, 38–41.

Boyer, P. D., Chance, B., Ernster, L., Mitchell, P., Racker, E. & Slater, E. C. (1977). *Annu. Rev. Biochem.* **46**, 955–1026.

Brand, M. D. (1977). *Biochem. Soc. Trans.* **5**, 1615–1620.

Brand, M. D., Reynafarje, B. & Lehninger, A. L. (1976). *Proc. Natl Acad. Sci. USA* **73**, 437–441.

Brand, M. D., Harper, W. G., Nicholls, D. G. & Ingledew, W. J. (1978). *FEBS Lett.* **95**, 125–129.

Bygrave, F. L. (1977). *Curr. Topics Bioenerg.* **6**, 259–318.

Carafoli, E. & Crompton, M. (1977). *Curr. Topics Membr. Trans.* **10**, 151–216.

Chance, B. & Williams, G. R. (1955). *J. Biol. Chem.* **217**, 409–427.

Chappell, J. B. (1968). *Brit. Med. Bull.* **24**, 150–157.

Chappell, J. B. (1977). "ATP", Carolina Biology Reader 50. Carolina Biological Supply Co., Burlington, N. Carolina, USA.

Chappell, J. B. (1979). *Trends Biochem. Sci.* **4**, N3–N4.

Chappell, J. B. & Crofts, A. R. (1966). *In* "Regulation of Metabolic Processes in Mitochondria" (J. M. Tager, S. Papa, E. Quagliariello & E. C. Slater, ed.) pp. 293–314. Elsevier, Amsterdam.

Chappell, J. B. & Haarhoff, K. N. (1967). *In* "Biochemistry of Mitochondria" (E. C. Slater, Z. Kaniuga & L. Wojtchak, ed.) pp. 75–91. Academic Press, London & New York.

Clayton, R. K. & Sistrom, W. R. (1979). "The Photosynthetic Bacteria". Plenum, New York.

Crofts, A. R. & Wood, P. M. (1977). *Curr. Topics Bioenergetics* **7**, 175–244.

Crompton, M. & Carafoli, E. (1979). *Methods Enzymol.* **56**, 338–352.

Crompton, M., Capano, M. & Carafoli, E. (1976). *Eur. J. Biochem.* **69**, 453–462.

Dawson, A. G. (1979). *Trends Biochem. Sci.* **4**, 171–176.

DePierre, J. W. & Ernster, L. (1977). *Annu. Rev. Biochem.* **46**, 201–262.

Downie, J. A., Gibson, F. & Cox, G. B. (1979). *Annu. Rev. Biochem.* **48**, 103–132.

Drachev, L. A., Jasaitis, A. A., Kaulen, A. D., Kondrashin, A. A., Liberman, E. A., Nemecek, I. B., Ostroumov, S. A., Semenov, A. Y. & Skulachev, V. P. (1974). *Nature (London)* **249**, 321–324.

Dutton, P. L. (1978). *Methods Enzymol.* **54**, 411–435.

Dutton, P. L. & Prince, R. C. (1978) *In* "The Photosynthetic Bacteria" (R. K. Clayton & W. R. Sistrom, ed.) pp. 525–570, Plenum, New York.

Dutton, P. L. & Wilson, D. F. (1974). *Biochim. Biophys. Acta* **346**, 165–212.

Eisenbach, M. & Caplan, S. R. (1977). *Trends Biochem. Sci.* **2**, 245–247.

Eisenbach, M. & Caplan, S. R. (1979). *Curr. Topics Membr. Trans.* **12**, 165–248.

Erecinska, M., Veech, R. L. & Wilson, D. F. (1974). *Arch. Biochem. Biophys.* **160**, 412–421.

Ernster, L. & Lee, C. P. (1964). *Annu. Rev. Biochem.* **33**, 729–788.

Ferguson, S. J., Jones, O. T. G., Kell, D. B. & Sorgato, M. C. (1979). *Biochem. J.* **180**, 75–85.

Fillingame, R. H. (1980). *Annu. Rev. Biochem.* **49**, 1079–1113.

Fonyo, A., Ligeti, E., Palmieri, F. & Quagliariello, E. (1975). *In* "Biomembranes, Structure and Function" (G. Gardos & I. Szasz. ed.) pp. 287–306. North-Holland, Elsevier.

Garland, P. B. (1978). *Nature (London)* **276**, 8–9.

Gibson, F., Cox, G. B. & Downie, J. A. (1979). *Trends Biochem. Sci.* **4**, 260–263.

Gingras, G. (1978). *In* "The Photosynthetic Bacteria" (R. K. Clayton & W. R. Sistrom, ed.) pp. 119–131. Plenum, New York.

Gomez-Poyou, A. & Gomez-Lojero, C. (1977). *Curr. Topics Bioenerg.* **6**, 221–257.

Greville, G. D. (1969). *Curr. Topics Bioenerg.* **3**, 1–78.

Gromet-Elhanan, Z. (1977). *Trends Biochem. Sci.* **2**, 275–277.

Haddock, B. A. (1980). *Phil. Trans. R. Soc. Lond. B* **290**, 329–339.

Haddock, B. A. & Jones, C. W. (1977). *Bact. Rev.* **41**, 47–99.

Hamilton, W. A. (1975). *Advan. Microbiol. Physiol.* **12**, 1–53.

Hamilton, W. A. (1977). *In* "Microbial Energetics" (B. A. Haddock & W. A. Hamilton, ed.) pp. 185–216. Cambridge University Press, Cambridge.

Harmon, H. J., Hall, J. D. & Crane, F. L. (1974). *Biochim. Biophys. Acta* **344**, 119–155.

Harold, F. M. (1977). *Curr. Topics Bioenerg.* **6**, 84–149.

Hatefi, Y. (1978). *Methods Enzymol.* **53**, 3–5 *et seq.*

Hatefi, Y., Haavik, A. G., Fowler, L. R. & Griffiths, D. E. (1962). *J. Biol. Chem.* **237**, 2661—2669.

Hauska, G. & Trebst, A. (1977). *Curr. Topics Bioenerg.* **6**, 151–220.

Henderson, P. J. F. (1971). *Annu. Rev. Microbiol.* **25**, 393–428.

Henderson, R. & Unwin, P. N. T. (1975). *Nature (London)* **257**, 28–32.

Hill, R. (1965). *Essays Biochem.* **1**, 121–151.

Hinkle, P. C. & Mitchell, P. (1970). *Bioenergetics* **1**, 45–60.

Höjeberg, B. & Rydström, J. (1979). *Methods Enzymol.* **55**, 275–283.

Ingledew, W. J., Cox, J. C. & Halling, P. J. (1977). *FEMS Microbiol. Lett.* **2**, 193–197.

Jagendorf, A. T. (1975). *In* "Bioenergetics of Photosynthesis" (Govindjee, ed.) pp. 413–492. Academic Press, New York & London.

Jagendorf, A. T. & Uribe, E. (1966). *Proc. Natl Acad. Sci. USA* **55**, 170–177.

John, P. & Whatley, F. R. (1977). *Biochim. Biophys. Acta* **463**, 129–153.

Jones, C. W. (1977). *In* "Microbial Energetics" (B. A. Haddock & W. A. Hamilton, ed.) pp. 23–59. Cambridge University Press, Cambridge.

Jones, O. T. G. (1977). *In* "Microbial Energetics" (B. A. Haddock & W. A. Hamilton, ed.) pp. 151–183. Cambridge University Press, Cambridge.

Junge, W. (1977). *Annu. Rev. Plant Physiol.* **28**, 503–536.

Junge, W. & Witt, H. T. (1968). *Z. Naturforsch. Teil B* **23**, 244–254.

Kaback, H. R. (1974). *Science* **186**, 882–892.

Kagawa, Y. (1972). *Biochim. Biophys. Acta* **265**, 297–338.

Kagawa, Y. (1978). *Biochim. Biophys. Acta* **505**, 45–93.

Kagawa, Y., Kandrach, A. & Racker, E. (1973). *J. Biol. Chem.* **248**, 676–684.

Kagawa, Y., Sone, N., Hirata, H. & Yoshida, M. (1979). *Trends Biochem. Sci.* **4**, 31–33.

Kell, D. B. (1979). *Biochim. Biophys. Acta* **549**, 54–99.

Klingenberg, M. (1979a). *Trends Biochem. Sci.* **4**, 249–252.

Klingenberg, M. (1979b). *Methods Enzymol.* **56**, 229–232.

Klingenberg, M. & Rottenberg, H. (1977). *Eur. J. Biochem.* **73**, 125–130.

Konings, W. N. (1979). *Methods Enzymol.* **56**, 378–388.

Kozlov, I. A. & Skulachev, V. P. (1977). *Biochim. Biophys. Acta* **463**, 29–89.

Kröger, A. (1978). *Biochim. Biophys. Acta* **505**, 129–145.

Kröger, A. & Klingenberg, M. (1973). *Eur. J. Biochem.* **34**, 358–365.

LaNoue, K. F. & Schoolworth, A. C. (1979). *Annu. Rev. Biochem.* **48**, 871–922.

Lee, C.-P. (1979). *Methods Enzymol.* **55**, 105–114.

Lundegårdh, H. (1945). *Arch. Bot.* **32A**, 12, 1.

Meijer, A. J. & van Dam, K. (1974). *Biochim. Biophys. Acta* **346**, 213–244.

McCarty, R. E. (1979). *Trends Biochem. Sci.* **4**, 28–30.

Mitchell, P. (1961). *Nature (London)* **191**, 423–427.

Mitchell, P. (1962). *Biochem. Soc. Symp.* **22**, 142–168.

Mitchell, P. (1966). "Chemiosmotic Coupling in Oxidative and Photosynthetic Phosphorylation". Glynn Research, Bodmin, Cornwall, U.K.

Mitchell, P. (1968). "Chemiosmotic Coupling and Energy Transduction". Glynn Research, Bodmin, Cornwall, U.K.

Mitchell, P. (1974a). *FEBS Lett.* **33**, 267–274.

Mitchell, P. (1974b). *FEBS Lett.* **43**, 189–194.

Mitchell, P. (1975a). *FEBS Lett.* **56**, 1–6.

Mitchell, P. (1975b). *FEBS Lett.* **59**, 137–139.

Mitchell, P. (1976a). *J. Theoret. Biol.* **62**, 327–367.

Mitchell, P. (1976b). *Biochem. Soc. Trans.* **4**, 399–430.

Mitchell, P. (1977). *FEBS Lett.* **78**, 1–20.

Mitchell, P. (1979a). *Eur. J. Biochem.* **95**, 1–20.

Mitchell, P. (1979b). *Science* **206**, 1148–1159.

Mitchell, P. & Moyle, J. (1958). *Proc. Roy. Soc., Edinburgh* **27**, 61–72.

Mitchell, P. & Moyle, J. (1965). *Nature (London)* **208**, 147–151.

Mitchell, P. & Moyle, J. (1967a). *Biochem. J.* **105**, 1147–1162.

Mitchell, P. & Moyle, J. (1967b). *Biochem. J.* **104**, 588–600.

Mitchell, P. & Moyle, J. (1968). *Eur. J. Biochem.* **4**, 530–539.

Mitchell, P. & Moyle, J. (1969a) *Eur. J. Biochem.* **7**, 471–484.

Mitchell, P. & Moyle, J. (1969b). *Eur. J. Biochem.* **9**, 149–155.

Mitchell, P., Moyle, J. & Mitchell, R. (1979). *Methods Enzymol.* **55**, 627–640.

Moore, A. L. & Rich, P. R. (1980). *Trends Biochem. Sci.* **5**, 284–288.

Moyle, J. & Mitchell, P. (1973). *FEBS Lett.* **30**, 317–320.

Moyle, J. & Mitchell, P. (1978a). *FEBS Lett.* **90**, 361–365.

Moyle, J. & Mitchell, P. (1978b). *FEBS Lett.* **88**, 268–272.

Munn, E. A. (1974). "The Structure of Mitochondria". Academic Press, London & New York.

Nedergaard, J. & Cannon, B. (1979). *Methods Enzymol.* **55**, 3–28.

Nelson, N. (1976). *Biochim. Biophys. Acta* **456**, 314–338.

Neumann, J. & Jagendorf, A. T. (1964). *Arch. Biochem. Biophys.* **107**, 109–119.

Nicholls, D. G. (1974). *Eur. J. Biochem.* **50**, 305–315.

Nicholls, D. G. (1976). *Trends Biochem. Sci.* **1**, 128–130.

Nicholls, D. G. (1979). *Biochim. Biophys. Acta* **549**, 1–29.

Nicholls, D. G. & Bernson, V. S. M. (1977). *Eur. J. Biochem.* **75**, 601–612.

Nicholls, D. G. & Crompton, M. (1980). *FEBS Lett.* **111**, 261–268.

Okamotu, H., Sone, N., Hirata, H., Yoshida, M. & Kagawa, Y. (1977). *J. Biol. Chem.* **252**, 6125–6131.

Ovchinnikov, Y. A. (1979). *FEBS Lett.* **94**, 321–336.

Palmer, J. M. (1979). *Biochem. Soc. Trans.* **7**, 246–252.

Palmieri, F. & Klingenberg, M. (1979). *Methods Enzymol.* **56**, 279–301.

Papa, S. (1976). *Biochim. Biophys. Acta* **456**, 39–84.

Park, R. B. & Sane, P. V. (1971). *Annu. Rev. Plant Physiol.* **22**, 395–430.

Parson, W. W. & Cogdell, R. J. (1975). *Biochim. Biophys. Acta* **416**, 105–149.

Pedersen, P. L. (1975). *Bioenergetics* **6**, 243–275.

Penefsky, H. S. (1979). *Adv. Enzymol. Relat. Areas Mol. Biol.* **49**, 223–280.

Pressman, B. C. (1976). *Annu. Rev. Biochem.* **45**, 501–530.

Prince, R. C. & Dutton, P. L. (1978). *In* "The Photosynthetic Bacteria" (R. K. Clayton & W. R. Sistrom, ed.) pp. 439–453. Plenum, New York.

Puskin, J. S., Gunter, T. E., Gunter, K. K. & Russell, P. R. (1976). *Biochemistry* **13**, 4811–4817.

Racker, E. (1975). *Biochem. Soc. Trans.* **3**, 785–802.

Racker, E. (1976). *Trends Biochem. Sci.* **1**, 244–247.

Racker, E. (1979). *Methods Enzymol.* **55**, 699–711.

Racker, E. & Stoekenius, W. (1974). *J. Biol. Chem.* **249**, 662–663.

Ragan, C. I. (1976). *Biochim. Biophys. Acta* **456**, 249–290.

Ragan, C. I. & Heron, C. (1978). *Biochem. J.* **174**, 783–790.

Ramos, S., Schuldiner, S. & Kaback, H. R. (1979). *Methods Enzymol.* **55**, 680–688.

Reeves, S. G. & Hall, D. O. (1978). *Biochim. Biophys. Acta* **463**, 275–297.

Reynafarje, B., Brand, M. D., Alexandre, A. & Lehninger, A. L. (1979). *Methods Enzymol.* **55**, 640–656.

Rieske, J. S. (1971). *Arch. Biochem. Biophys.* **145**, 179–193.

Rieske, J. S. (1976). *Biochim. Biophys. Acta* **456**, 195–247.

Rosing, J. & Slater, E. C. (1972). *Biochim. Biophys. Acta* **267**, 275–290.

Rottenberg, H. (1975). *Bioenergetics* **7**, 61–74.

Rottenberg, H. (1976). *FEBS Lett.* **66**, 159–163.

Rottenberg, H. (1979a). *Methods Enzymol.* **55**, 547–569.

Rottenberg, H. (1979b). *Biochim. Biophys. Acta* **549**, 225–253.

Rydström, J. (1977). *Biochim. Biophys. Acta* **463**, 155–184.

Ryrie, I. J. & Jagendorf, A. T. (1972). *J. Biol. Chem.* **24**, 4453–4459.

Saris, N.-E., L. & Åkerman, K. E. O. (1980). *Curr. Topics Bioenergetics* **10**, 103–179.

Scarpa, A. (1979). *Methods Enzymol.* **56**, 301–338.

Scheurich, P., Schäfer, H.-J. & Dose, K. (1978). *Eur. J. Biochem.* **88**, 253–257.

Selwyn, M. J., Dawson, A. P. & Dunnett, S. J. (1970). *FEBS Lett.* **10**, 1–5.

Senior, A. E. (1973). *Biochim. Biophys. Acta* **301**, 249–277.

Shavit, N. (1980). *Annu. Rev. Biochem.* **49**, 111–138.

Singer, S. J. & Nicolson, G. L. (1972). *Science* **175**, 720–731.

Skulachev, V. P. (1970). *Biochim. Biophys. Acta* **216**, 30–42.

Skulachev, V. P. (1971). *Curr. Topics Bioenergetics* **4**, 127–190.

Skulachev, V. P. (1976). *FEBS Lett.* **64**, 23–25.

Skulachev, V. P. (1979). *Methods Enzymol.* **55**, 586–603.

Slater, E. C. (1953). *Nature (London)* **172**, 975–976.

Slater, E. C. (1974). *Biochem. Soc. Trans.* **2**, 1149–1163.

Slater, E. C., Rosing, J. & Mol, A. (1973). *Biochim. Biophys. Acta* **292**, 534–553.

Sone, N., Yoshida, M., Hirata, H. & Kagawa, Y. (1977). *J. Biol. Chem.* **252**, 2956–2960.

Sorgato, M. C., Ferguson, S. J., Kell, D. B. & John, P. (1978). *Biochem. J.* **174**, 237–256.

Stoekenius, W. (1976). *Sci. Amer.* **234**, 38–46.

Stoekenius, W., Lozier, R. H. & Bogomolni, R. A. (1979). *Biochim. Biophys. Acta* **505**, 215–278.

Tedeschi, H. (1979). *Trends Biochem. Sci.* **4**, N182–N185.

Thayer, W. S. & Hinkle, P. C. (1973). *J. Biol. Chem.* **248**, 5395–5402.

Thayer, W. S. & Hinkle, P. C. (1975). *J. Biol. Chem.* **250**, 5330–5335.

Trebst, A. (1976). *Trends Biochem. Sci.* **1**, 60–62.

Unwin, P. N. T. & Henderson, R. (1978). *J. Mol. Biol.* **94**, 425–440.

Van Dam, K., Westerhoff, H. V., Krab, K., van der Meer, R. & Arents, J. C. (1980). *Biochim. Biophys. Acta* **591**, 240–250.

Vignais, P. V. (1976). *Biochim. Biophys. Acta* **456**, 1–38.

Von Jagow, G., Schägger, H., Engel, W. D., Machleidt, W., Machleidt, I. & Kolb, H. J. (1978). *FEBS Lett.* **91**, 121–125.

Waggoner, A. S. (1976). *J. Mem. Biol.* **27**, 317–334.

Walz, D. (1979). *Biochim. Biophys. Acta* **505**, 279–354.

West, I. C. & Mitchell, P. (1972). *J. Bioenergetics* **3**, 445–462.

West, I. C. & Mitchell, P. (1973). *Biochem. J.* **132**, 587–592.

Westerhoff, H. V. & van Dam, K. (1979). *Curr. Topics Bioenergetics* **9**, 1–62.

Wikström, M. K. F. (1973). *Biochim. Biophys. Acta* **301**, 155–193.

Wikström, M. K. F. & Krab, K. (1978). *FEBS Lett.* **91**, 8–14.

Wikström, M. K. F. & Krab, K. (1979). *Biochim. Biophys. Acta* **549**, 177–222.

Wikström, M. K. F. & Krab, K. (1980). *Curr. Topics Bioenergetics* **10**, 51–101.

Wilbrandt, W. (1974). *In* "Biomembranes" (H. Eisenberg *et al.*, ed.) Vol. 7, 11–31. Plenum, New York.

Williams, R. J. P. (1961). *J. Theor. Biol.* **1**, 1–13.

Williams, R. J. P. (1976). *Trends Biochem. Sci.* **1**, N222–N224.

Williams, R. J. P. (1978a). *Biochim. Biophys. Acta* **505**, 1–44.

Williams, R. J. P. (1978b). *FEBS Lett.* **85**, 9–19.

Wilson, D. B. (1978). *Annu. Rev. Biochem.* **47**, 933–965.

Witt, H. T. (1979). *Biochim. Biophys. Acta* **505**, 355–427.

Wraight, C. A., Cogdell, R. J. & Chance, B. (1978). *In* "The Photosynthetic Bacteria" (R. K. Clayton & W. R. Sistrom, ed.) pp. 471–511. Plenum, New York.

Zalman, L. S., Nikaido, H. & Kagawa, Y. (1980). *J. Biol. Chem.* **255**, 1771–1774.

Zilberstein, D., Schuldiner, S. & Padan, E. (1979). *Biochemistry* **18**, 669–673.

Subject Index

A23187, 33
Acetate, mitochondrial permeability, 36–39
Acid-bath experiment, 95–96
Active transport, 28
Adenine nucleotide translocator, 88, 159–164
  effect of membrane potential, 163
  mechanism, 162–163
ADP/O ratio, 87, 91–93
Affinity, 62
9-aminoacridine, 74, 146
"Ammonium swelling", 167–169
Antennae, 133–135, 143
Antimycin A, 94, 110, 120–121, 125, 127, 140
Antiport, 27
  electrochemical potential equation for, 55
Arsenazo III, 173
Arum maculatum, 127
Ascorbate, 109, 110, 122
ATPase, *see* ATP synthetase
ATP synthesis
  by artificial proton electrochemical potential, 95
  in reconstituted systems, 96
  thermodynamics, 45–48
ATP synthetase, 151–164
  adenine nucleotide binding, 158–159
  chloroplast, 146
  conformational changes, 157–159
  coupling to proton electrochemical potential, 3–5
  of *E. coli*, 153
  H^+/ATP stoicheiometry, 61, 82–84
  mechanism, 155–160
  orientation, 151

proton translocation, 21
reconstitution, 117
structure, 151–155
thermodynamics of reaction, 45–48
thermophilic bacteria, 152–159
vectorial group translocation model, 18–20
Atractylate, 86, 88
Atractylis gummifera, 161
Atractyloside, *see* atractylate
Azide, 123

Bacterial transport, 175–177
Bacteriochlorophyll, 135–138
Bacteriopheophytin, 135–138
Bacteriorhodopsin, 22, 117, 147–149
bc_1 Complex, *see* Complex III
Bongkrekic acid, 161–164
Brown adipose tissue mitochondria, 39, 83, 90
N-butylmalonate, 86

Calcium and plant mitochondria, 127
Calcium transport by mitochondria, 172–175
  electrochemical equations, 55
  evidence for uniporter, 38, 172
  independent efflux pathways, 172–175
  role of permeant anions, 77, 174
Carbon monoxide, 123
Carbonyl cyanide-*p*-trifluoromethoxy-phenyl-hydrazone, *see* FCCP
Carnitine carrier, 169–170
Carotenoid band shift, 73, 138–145
CF_1-ATPase, 152–159
Chemical coupling, 13–22

Chemiosmotic hypothesis
  "central dogma", 3–5, 12–22
  experimental verification, 20–22
Chemolithotrophic bacteria, 129–130
Chlorophyll, 138, 142–145
Chloroplasts
  electron-transfer inhibitors, 143
  non-cyclic electron transfer, 133, 142–146
  preparation, 9–10
  structure, 7–10
  values for proton electrochemical potential, 74
Cholate dialysis, 12, 115
Chromatophores, 135–142
  values for proton electrochemical potential, 74
Coenzyme Q, see Ubiquinone
Complex I, 117–119
  $H^+/2e^-$ stoicheiometry, 81
  reversed electron transfer, 93–96
Complex II, 118–119
  $H^+/2e^-$ stoicheiometry, 81
Complex III, 119–122
  $H^+/2e^-$ stoicheiometry, 81
  redox potentiometry, 112
Complex IV see Cytochrome c oxidase
Conformational hypothesis, 18
"Conformational pump", 80, 114–115, 122–124
Copper in respiratory chain, 103, 122–124
CoQ see Ubiquinone
Co-transport, see Symport
Cyanide, 86, 110, 122–124, 126
Cyanide-resistant oxidation, 127
Cyanine dyes, 75
Cyanohydroxycinnamate, 169
Cyclic electron transfer, 133, 138–142
Cyclic photophosphorylation, 145
Cytochromes, 103–105
  spectra, 103–104
  liquid N_2 spectra, 104–105
Cytochrome a, 103
Cytochrome a_3
  reaction with CO, 105
Cytochrome b, 103, 119–122, 127–130, 137–142
  in Complex II, 118
  resolution by redox potentiometry, 111, 119–123

Cytochrome b_K, see Cytochrome b_{566}
Cytochrome b_T, see Cytochrome b_{562}
Cytochrome b_{562}, 105, 119–122
Cytochrome b_{566}, 105, 119–122
Cytochrome $b_{556}^{NO_3}$, 129
Cytochrome c, 49, 103, 105, 122
  spectra, 104
Cytochrome c oxidase, H^+/O stoicheiometry, 81
Cytochrome c_1, 105, 120
Cytochrome c_2, 137
Cytochrome d, 103, 128
Cytochrome f, 143
Cytochrome o, 103, 128

Deoxycholate, 115
Dicarboxylate carrier, 170
Dicyclohexylcarbodiimide, 153–154
Difference spectra, 102–104
Diffusion potentials, 58, 75
Dithionite, 85, 110, 119
Donnan potentials, 58

E. coli
  respiratory chains, 127–130
  values for proton electrochemical potential, 74
Electrical transport, 27–28
Electrically permeant ions
  effect on $\Delta\Psi$, 76
  and osmotic swelling, 36–39
Electrochromism, 73, 138–145
Electron paramagnetic resonance, see Electron spin resonance
Electron spin resonance, 106–107, 121
Electron-transfer chain, see Respiratory chain
Electron-transferring flavoprotein, 100, 118
Electron-transport particles, see sub-mitochondrial particles
Electroneutral transport, 27–28
  effect on pH gradient, 77
Electrophoretic transport, 27
Energy-transducing intermediate, 12–22
Energy-transducing membranes, see also individual organelles, 1–12
  electrical field across, 30
  structure, 25–26

Energy transduction, pathways, 14
Enthalpy, 42
Entropy, 42
EPR, see Electron spin resonance
Equilibrium, constant, 44–48
Equilibrium
  definition, 43–44
Equilibrium distributions across mem-
    branes, 56–58
  as function of membrane potential and
    charge, 56
ESR, see Electron spin resonance
N-ethylmaleimide, 80, 157, 159
Exchange carriers, see Transport proteins
Exchange-diffusion, see Antiport

FAD, 100, 103, 118, 119, 129
FCCP
  mechanism, 34
  and membrane potential, 72, 76
  and mitochondrial swelling, 38
  and proton extrusion, 79
  and respiratory control, 86, 90–93
  structure, 34
Ferredoxin, 107, 142–146
Ferredoxin-NADP reductase, 143–145
Ferricyanide, 109, 111, 118, 121, 145
Ferrocyanide, 125
Fe/S-proteins, see Iron–sulphur proteins
Flavin adenine dinucleotide, see FAD
Flavin mononucleotide, see FMN
Flavoproteins, 103
Flow dialysis, 71–72
Fluid-mosaic model, 25–26
FMN, 100, 103, 114, 117
F_0, 154
  proton conduction, 155, 156
Formate dehydrogenase, 128–129
Free energy, see Gibbs energy
F_1-ATPase, 152–159

Gibbs energy, 41–63
  biochemical conventions, 47–48
  standard change, 46
Glucose–hexokinase trap, 92–93
Glutamate–aspartate carrier, 170
Glutamate–aspartate cycle, 171
Glutamate carrier, 170

α-Glycerophosphate dehydrogenase, 100,
    118
α-Glycerophosphate shuttle, 171–172
Gramicidin, 22, 31–32
Grana, 9
Group translocation, see Vectorial group
    translocation

Half reaction, 49
Halobacteria, 147–149
Halobacterium halobium, 147–149
High-energy bonds, 48
High-energy intermediate, see Energy-
    transducing intermediate
High-potential iron–sulphur protein, 107
"Hydrogen" transfer, 49
Hydrogenase, 128–129

Ion electrochemical potential differences,
    53–55
  relation to Gibbs energy changes, 54
Ionophores, see also individual compounds,
    30–35
  channel forming, 30–34
  mobile carriers, 30–34
Ion transport, 26–39
  across bilayers, 29–30
Iron–sulphur proteins, 103–107, 118, 120,
    127–130
  detection by electron spin resonance,
    106
  in purple bacteria, 137–142
  structure, 105, 107
Irreversible thermodynamics, 62–63

Kabackosomes, 7

lac permease, 175–177
Light scattering and mitochondrial trans-
    port, 36–39, 167–169
Lipid bilayers
  formation, 11–12
  permeability properties, 29–30
Lipophilic ions, 21, 34–35, 124–126, 144
Liposomes, 11–12
Looped respiratory chains, 80, 114, 122,
    127

Malonate, 86
Mass–action ratio, 44–48
Membrane potential
  in chloroplasts, 146
  generation by ATP hydrolysis, 76
  generation by diffusion potential, 76
  generation by respiratory chain, 76
  by ion-specific electrodes, 69–71
  by isotope distribution, 71–73
  by lipophilic ions, 69, 74
  and Nernst equation, 56
  by optical indicators, 73–75
  and proton electrochemical potential,
    54
Menaquinone, 107–108, 128, 129
p-Mercuribenzoate, 159
Mersalyl, 159
Metabolite transport, 167–177
  methods, 167–169
Methyltriphenylphosphonium, see TPMP
Mid-point potential, 51–52, 113, 119–122,
    126
  in chloroplasts, 144
  of mitochondrial redox components,
    109–113
  in reaction centres, 135–145
Mitochondria
  cristae, 6
  matrix buffering capacity, 70, 76
  metabolite carriers, 167–172
  outer membrane, 6
  preparation, 7
  structure, 5–7
  values for proton electrochemical
    potential, 69, 74, 83, 89, 90

NAD$^+$
  reduction, 49, 99–100, 124
  structure, 100
NADH
  oxidation by mitochondria, 170–172
  oxidation by plant mitochondria, 127
  as substrate for complex I, 109, 117–118
NADH-cytochrome c oxidoreductose,
    117
NADH-dehydrogenase, 117, 128, 129
NADH-UQ oxidoreductase, see Complex
    I
Nernst equation, 56, 73

Neutral Red, 146
Nicotinamide adenine dinucleotide, see
    NAD$^+$
Nicotinamide nucleotide transhydro-
    genase, 124, 126
Nigericin, 32–33, 37–38
Non-equilibrium thermodynamics, 62–63
Non-haem iron proteins, see Iron–sulphur
    proteins

Oligomycin, 13, 76, 82, 86, 88, 93, 94,
    153–155
Oligomycin-sensitivity-conferring pro-
    tein, 153
Ornithine carrier, 170
Osmotic swelling, 36–39, 77–78, 167–169
Oxido-reduction potentials, see Redox
    potentials
2-Oxoglutarate carrier, 170
Oxonols, 75
Oxygen electrode, 84–91
  and ADP/O ratios, 92
  and mitochondrial respiratory chain,
    109–110
Oxygen pulse technique, 78–80

Paracoccus denitrificans, 128
pH gradient
  in chloroplasts, 146
  by ion specific electrodes, 69–71
  by isotope distribution, 71–73
  by optical indicators, 73–75
  and proton electrochemical potential,
    54
Phenazine methosulphate, 145
Phenyldicarbaundecaborane, 126
Phosphate carrier, 159–164
Phosphate potential, 48
Phosphate transport, effect on H$^+$/O
    stoicheiometry, 78–79
Phosphorylation potential, 48, 83, 88
Phosphotransferase system, 177
Photons, 59–60
Photosynthesis, 133–146
Photosystem I, 142–146
Photosystem II, 142–146
Piericidin A, 118
Planck's constant, 59

Plastocyanine, 143–146
Plastoquinone, 107–108, 142–146
P/O ratio, 91–93
Potassium
  ionophore-induced permeability, 30–33, 37–39
  mitochondrial permeability, 36–39
Protein-catalysed transport, 35–36
Proton circuit, 65–96
  analogy to electrical circuit, 4–5, 65–67, 85–86
  in chloroplasts, 146
  localized microcircuits, 19–22
Proton conductance, 21, 89–91
  and proton translocators, 33
Proton current, 78, 84, 88–91
Proton electrochemical potential
  derivation of equation, 54
  effect of ADP, 88–90
  effect of FCCP, 90
  effect of surface potential, 59
  as energy coupling intermediate, 3
  examples of organelles, 74
  generation in purple bacteria, 138–142
  by ion-specific electrodes, 69–71
  by isotope distribution, 71–73, 83, 89, 90
  measurement, 68–75
  relative contributions of $\Delta\Psi$ and ΔpH, 75–78, 96
Protonmotive force, see Proton electrochemical potential
Protonmotive Q-cycle, see Q-cycle
Proton translocation
  by ATP synthetase, 20–22
  by photosynthetic electron-transfer chain, 20–22
  by respiratory chain, 20–21
Proton translocators, see also specific compounds, 13, 33–34, 82, 88, 125
  effect on proton circuit, 66
Purple membrane, 147–149
Pyruvate, transport into mitochondria, 36, 169, 170

Q, see Ubiquinone
Q-cycle, 107, 119–122
Quanta, 59

Reaction centres, 10, 133–145
  electron transfer, 136
  organization in membrane, 139, 145
  structure, 135–137
Reconstituted systems, 10–12, 117
Redox couple, 49–51, 99
Redox mediator, 50, 109
Redox potentials, 43, 49–53, 60–61, 109–113
  in mitochondrial respiratory chain, 99–113
  relation to Gibbs energy changes, 53
  variation with extent of reduction, 52
Redox potentiometry, 109–113
Respiratory chain
  in aerobic Rps. sphaeroides, 141–142
  of bacteria, 127–130
  components, 99–108
  fractionation and reconstitution, 115–117
  linear sequence, 108–113
  "looped" electron-transfer, 16–22
  methods of detection, 99–108
  mitochondrial, 100, 108–126
  of plant mitochondria, 126–127
  proton pumps, 67
Respiratory control, 87, 89, 91
  and proton electrochemical potential, 87–91
Respiratory states, 69
Retinal, 147–149
Reversed electron transfer, 93–95
  in photosynthetic bacteria, 140–142
Rhodopseudomonas sphaeroides, 136–142
Rieske protein, 120
Rotenone, 110, 118
Ruthenium Red, 174

Safranine, 75, 76
Semiquinones, 121, 136–137, 143
Silicone oil centrifugation, 71–73
Spectrophotometer
  dual-wavelength, 100–105
  split-beam, 100–105, 111–112
  stopped-flow, 100–105
Spectroscopy, 101–105
Sphaeroplasts, 7
"Squiggle", see Energy-transducing intermediate

Standard hydrogen electrode, 50
Standard redox potential, 50–53
  pH dependency, 51
State 4 respiration, and proton conduc-
    tance, 91
State 4–state 3 transition, 89
Stoicheiometry
  ADP/2e$^-$, 84
  H$^+$/ATP, 61
  H$^+$/2e$^-$, 61, 78–82, 93, 114–115, 122–
    124, 146
  charge stoicheiometry, 80–81, 122–124
Sub-mitochondrial particles
  cytochrome spectra, 104
  H$^+$/ATP stoicheiometry, 82, 93
  structure, 5–7
  values for proton electrochemical
    potential, 74
Substrate-level phosphorylation, 13–15
Succinate dehydrogenase, 100, 118, 119
Sucrose-impermeable space, 73
Surface potentials, 58
Symport, 27

Tetramethyl-p-phenylenediamine, see
    TMPD
Tetraphenylborate, 35
TF$_1$-ATPase, 152–159
Thermodynamic systems, 41–42
Thermophilic bacterium PS 3, 152–159
Thiobacillus ferro-oxidans, 130
Thiocyanate
  as membrane potential indicator
  mitochondrial permeability, 36–38
Thylakoid membranes, 8–10, 142–146
Thylakoid space, 8

TMPD, 109, 110, 122–124
TPMP, 34–35, 78
  structure, 35
Transport proteins, 21–22
Tricarboxylate carrier, 170
Two-electron gate, 137

Ubiquinol, see also Ubiquinone, 107–108
Ubiquinone, 107–108, 119
  as mobile redox carrier, 107, 127
  reduction, 49
  role in complex III, 107, 119–122
  role in reaction centres, 135–145
Uncontrolled respiration, 89
Uncouplers, see Proton translocators
Uncoupler-stimulated ATPase, 93
Uniport, 27
  electrochemical potential equation for,
    55
UQ, see Ubiquinone
UQ-cytochrome c oxidoreductase, see
    Complex III

Valinomycin, 30–33, 36–38, 58, 69, 70,
    75–80, 125, 139, 140, 156
Vectorial group translocation, 15–22, 29,
    80–82, 175–177

Weak acids
  equilibrium distribution, 56–58
  to determine ΔpH, 56–58
Weak bases
  equilibrium distribution, 56–58
  to determine ΔpH, 56–58